Landscape as a Geosystem

László Miklós · Erika Kočická
Zita Izakovičová · Dušan Kočický
Anna Špinerová · Andrea Diviaková
Viktória Miklósová

Landscape as a Geosystem

Springer

László Miklós
UNESCO-Chair for Sustainable Development
Technical University in Zvolen
Zvolen, Slovakia

Anna Špinerová
UNESCO-Chair for Sustainable Development
Technical University in Zvolen
Zvolen, Slovakia

Erika Kočická
UNESCO-Chair for Sustainable Development
Technical University in Zvolen
Zvolen, Slovakia

Andrea Diviaková
UNESCO-Chair for Sustainable Development
Technical University in Zvolen
Zvolen, Slovakia

Zita Izakovičová
Institute of Landscape Ecology SAS
Bratislava, Slovakia

Viktória Miklósová
Institute of Landscape Ecology SAS
Bratislava, Slovakia

Dušan Kočický
ESPRIT Ltd.
Banská Štiavnica, Slovakia

ISBN 978-3-030-06775-5 ISBN 978-3-319-94024-3 (eBook)
https://doi.org/10.1007/978-3-319-94024-3

Softcover re-print of the Hardcover 1st edition 2019

Printed on acid-free paper

This Springer imprint is published by the registered company Springer Nature Switzerland AG
The registered company address is: Gewerbestrasse 11, 6330 Cham, Switzerland

Acknowledgement

The publication is the result of the research supported by the grant agency KEGA Project No. 013TU Z-4/2016 and by grant agency VEGA Project No. 1/0096/1614-0735 and VEGA Project No. 2/0066/15.

Contents

Abbreviations

ABC	Abiotic complex (abiocomplex)
B.p.v.	Baltic vertical reference system
BAC	Biotic–anthropic complex
BM	Basic map
BSEU	Bonited Soil-Ecological Unit
CLS	Current landscape structure
CSD	Commission for Sustainable Development
CSFR	Czechoslovak Federal Republic
DTM	Digital Terrain Model
EC	European Commission
EIA	Environmental Impact Assessment
ELC	European Landscape Convention
EU	European Union
FAR	Folk Architecture Reservation
FH	Forest habitats
FMP	Forest Management Plan (recently FCP—Forest Care Programme)
FMU	Forest Management Unit
GEOIS	Geological information system
GIB-GES	Complex geological information base for the needs of protecting nature and landscape management
GIS	Geographic Information Systems
HGT	Hydrogeological type
HNC	Hydrological number of catchment
HPZ/SZ	Hygienic Protection Zone/Safety (buffer) Zone
IFM	Integrated flood management
IGEC	Integrated geocomplex
ILM	Integrate landscape management
INSPIRE	INfrastructure for SPatial InfoRmation in Europe
ISTB	Information system of taxons a habitats

KEGA	Cultural and Educational Grant Agency of the Ministry of Education, Science, Research and Sport of Slovak Republic
KPP	Complex Agricultural Soil Survey of Slovakia
LANDEP	LANDscape Ecological Planning
LEB ILM	Landscape ecological basis of integrated landscape management
LEC	Landscape ecological complex
LGIS	Forest geographical information system
LTSES	Local Territorial System of Ecological Stability
m a.s.l.	Metres above sea level
MAB	UNESCO's Man and the Biosphere Programme
MCR	Memorial City Reservation
MoE SR	Ministry of the Environment of the Slovak Republic
MPRV SR	Ministry of Agriculture and Rural Development of the Slovak Republic
MS	Microsoft
NATURA 2000	Network of Sites of Community Importance and Protected Bird Areas
NFC	National Forest Centre
NFH	Non-forest habitats
NISI	National Infrastructure of Spatial Information
NM	Nature Monument
NNM	National Nature Monument
NNR	National Nature Reservation
NP	National Park
NR	Nature Reservation
OECD	Organisation for Economic Co-operation and Development
PA	Protected Area
PBA	Protected Bird Area
PESD	Programme of Economic and Social Development
PGC	Physical–geographical complex
PLA	Protected Landscape Area
PLS	Primary landscape structure
PNV	Potential natural vegetation
POVAPSYS	Flood warning and prediction system
PT	Protected Tree
PWMA	Protected Water Management Area
PZ	Protection Zone
REPGES	REpresentative Potential GEoecoSystems (Research Institute for Soil Fertility)
SAS	Slovak Academy of Sciences
SCI	Site of Community Importance
SEA	Strategic Environmental Assessment
SEC	Socio-economic complex
SEF	Socio-economic factor (phenomenon)
SGIDŠ	State Geological Institute of Dionýz Štúr

S-JTSK	Coordinate system—Uniform trigonometric cadastral network
SkEA	Slovak Environmental Agency
SLS	Secondary landscape structure
SNC SR	State Nature Conservation of the Slovak Republic
SR	Slovak Republic
SSCRI	Soil Science and Conservation Research Institute (previously RISF—Research Institute for Soil Fertility)
STN	Slovak Technical Norm (standard)
TLS	Tertiary landscape structure
TSES	Territorial System of Ecological Stability
UNESCO	United Nations Educational, Scientific and Cultural Organization
USDF	Unit of Spatial Division of Forest
WRI	Water Research Institute
ZB GIS	Fundamental Database (of topographic objects) for Geographic Information

List of Figures

List of Tables

Introduction

Landscape is a common word, everybody knows what it is, there is no need to explain it. Is this statement true? Of course, we have no right to prevent anybody from having their own opinion about how to perceive the surrounding material reality in whatever way. On the other hand, we know that the landscape is a very complex entity influencing all our activities, and, in order to understand its functioning and its reactions to our activities, we need an amount of knowledge. Subsequently, when we have gained the knowledge, we wish to implement our scientific knowledge about nature and landscape into social practice.

The landscape is a very diverse entity, but at the same time it is a whole in its diversity, in which its individual components cannot exist without others. This view of diversity and holism at the same time might be addressed by a holistic approach. However, the holistic approach is insufficiently specific, and it permits very free interpretations of the landscape, most often as an image, as an ensemble of its visible elements, especially the elements of the land use, which forms only an aesthetic framework for human activities. What about the invisible elements of the landscape? Are the geological base, soils, georelief, ecosystems, climatic indicators and their interrelationships important for planning? Of course they are! In fact, they determine human activities more specifically. If we really want to define them globally, in complexes, in interrelationships, we need a systematic approach, which could be considered a specific concretization of the holistic approach. Such an approach resulted in the understanding of the landscape as a geosystem presented in this book.

It is to be emphasised that this approach has been motivated as mentioned in first sentences—in an effort to promote scientific knowledge of the landscape into practice. An essential aspect of this advance is the understanding of the landscape as a geosystem.

The landscape is the spatial setting and the material base for all the activities of humans. These activities meet in the same spatial–material entity, and therefore it is essential to regulate them. Regulation might have different bases but we consider that the principles of sustainable development undoubtedly prescribe the harmonisation of those activities with the given spatial–material entity as with the object where all the activities take place. This approach needs first of all a systematic *definition and description of this object*, as well as development of scientific methods transferring scientific knowledge to decision-making processes on the utilisation of this material basis for human development, through to multiple and multipurpose integrated landscape management.

The book defines this spatial setting and this material basis for human development as *landscape as a geosystem*. This book analyses the landscape as a geosystem in all its complexity (from the abiotic environment, through land use to its socio-economic character) as an integrated natural resource, as society's life space, as well as an object of planning and decision-making on sustainable land use. The landscape properties are presented in a form of databases usable for a variety of purposes relating to the Directive INSPIRE 2007/2/EC (INSPIRE—Infrastructure for Spatial InfoRmation in Europe) requirements to enable these databases to serve as a national spatial information database for the needs of applied landscape ecological researches and real spatial planning processes. The properties of the geosystem included in spatial information systems should then serve as the regulatives for the optimum spatial organisation of the activities in this setting as well as for the optimum mode of the utilisation of all points of this space.

In general, this book describes the landscape as a geosystem from a **purpose-oriented point of view**. The whole theory and methodology regarding the goals given by planning practice—to apply the scientifically defined material object to the legally supported planning processes.

Accordingly, the main themes of the book are as follows:

- a brief overview of the role of the landscape as a geosystem in planning, projecting and integrated landscape management;
- theoretical approaches to the landscape, describing different definitions from the physical–geographical definitions through geosystem theory up to the approach to the landscape as a picture;
- a definition of the landscape as a geosystem, its models, structures, elements and relations. This is the substantial part of the book;
- construction, assignation, delimitation and mapping of the geosystems, including the creation of the spatial information system;
- detailed characteristics of the partial geocomplexes, their elements and properties;
- application of the geosystems in a model territory—using the example of the Ipeľ basin (Slovak Republic).

Except for the theoretical–methodical approach to the geosystem, the other main objective of the book is to present a real applicable procedure for the creation a complex spatial database of the model territory as an objective, manifold usable, scientifically sound foundation for regional studies, programmes, planning, projecting and management of the sustainable organisation and utilisation of the landscape. The book presents an overview of map legends with complete domain values of selected attributes of all three landscape structures (primary, secondary and tertiary) routinely used in Slovakia. At the end, the publication presents an example of the construction and mapping of geocomplexes as well as the creation of the database on the model territory at the regional level. So, the book in its each part devotes a little bit broader attention to the players in landscape research in Central and Eastern Europe than it is usual in mainstream works in recent decades.

Chapter 1
The Material Base of Sustainable Development—The Landscape

Abstract The concept of the sustainable development requires an exact formulation of the main object of our interest—the **landscape**—in the sphere of science, politics, decision-making, planning and projecting. The different approaches to the landscape offer different possibilities for their implementation in legislation and real planning processes. The decisive step is the functional **integration of landscape-ecological knowledge** into existing management and planning processes. With simplification we can state that we need to integrate two groups of knowledge: the first and initial is the definition what actually is to be planned and managed. This is the "demand" from humans. The second one is the knowledge of the **landscape**, the landscape "offer, supply". The confrontation of these two groups should lead to a process referred to as the **ecologization of the landscape management**. The landscape is a comprehensive integrated spatial-material entity—the environment for the life of people and other organisms rather than just its separate components. At the same time it is the only space which we must all fit into. So, the landscape resources are used for different purposes for different activities that compete with each other and they cause mutual conflicts. Relevant environmental-political documents starting with Agenda 21 issued on Rio Summit 92, namely the Chapter 10 entitled "Integrated approach to the management of land resources" justified the need for the **integrated landscape management** as a major instrument of sustainable development. All abovementioned principles can be realized if the landscape is understood systematically, as **a geosystem**. This understanding has become the base for the legally defined and nowadays routine planning tools—the landscape-ecological planning LANDEP and projecting the territorial system of ecological stability TSES—as transmitters of the landscape-ecological knowledge into the spatial planning processes.

Keywords Landscape · Ecologisation · Integrated management · Plannings

The theory and practice of landscape sciences is decisively influenced by several basic concepts of sustainable development, such as environmental protection, management of natural resources, nature conservation, landscape design and planning procedures, integrated watershed management, and other policies. This development requires an exact formulation—or reformulation—of the main object of our inter-

L. Miklós et al., *Landscape as a Geosystem*,
https://doi.org/10.1007/978-3-319-94024-3_1

est—the **landscape**, as well as the most exact formulation possible, but at the same time acceptable in the sphere of policy, decision making, planning and projecting. The above-mentioned problems—as well as the diversity of approaches, their recent changes (Wu and Hobbs 2002; Haber 2002; Nassauer et al. 2007; Kienast et al. 2007; Nassauer and Opdam 2008; Mizgajski and Markuszewska 2010), even some kind of the "identity crisis" of landscape ecology, which according to Wu (2013) was perceived at the turn of the new millennium—force us repeatedly to reopen theoretical questions on definition of the landscape and its implementation in real policies.

The concept of the landscape occurs on broad scale of different sciences. Nowadays, basically at least two main streams should be identified: we may call these the "hard", geosystem-based concepts of the landscape, and the "soft", cultural heritage, "values" and perception-based ones. This division is not a ranking, just a differentiation. The first approach to the landscape is represented mainly by geographers and "geographically-biased" landscape ecologists, the second one by a very broad group of different specialists from other sciences, like social scientists, architects and artists. This group cannot be defined in a simple way but all its members might be considered as friends and lovers of the landscape. Without any doubt a landscape type defined exactly, e.g., by relief dissection, soil depth or a biocoenoses, has a different normative effect for the planning and management of the territory than the characteristics of the beauty of a landscape, which is much more subjective and changing, e.g., according to the persons, angle of view, etc.

These differences between the approaches are obvious with mutual comparisons of definitions based on general system theory von Bertalanffy (1968), modified for geographical and landscape sciences by the main representatives of the first group, such as Neef and others (Neef 1967; Neef et al. 1973); Chorley and Kennedy (1971); Preobrazhensky (Preobrazhensky and Minc 1973); Sochava (1977); Krcho (1968, 1978), with the definition given e.g. by the European Landscape Convention (see below). Of course, the above does not mean at all, that beauty and other similar characteristics or values of the landscape are not important indices of the landscape. The opposite is true: precisely because of their "fuzzy" character their implementation needs more diligence in order not to lose them in design and planning processes.

Nevertheless, those different approaches to the landscape offer different possibilities for their implementation in legislation and real planning processes. E.g., the geosystem approach has become the basis for implementation of landscape-ecological planning and eco-network planning in Slovakia. Recently, the cultural heritage concept is significantly supported by the European Landscape Convention.

1.1 Key Aspects of the Relation Geosystem versus Ecologization of Landscape Management

The theory and practice of landscape sciences decisively influence several basic concepts of sustainable development, such as environmental protection, management of natural resources, nature conservation, landscape design and planning procedures, integrated watershed management, and other policies. This development requires an exact formulation—or reformulation—of the main object of our interest—the **landscape**, in its most exact formulation possible while remaining acceptable in the sphere of policy, decision making, planning and projecting. The above-mentioned problems—as well as the diversity of approaches, their recent changes (Wu and Hobbs 2002; Haber 2002; Nassauer et al. 2007; Kienast et al. 2007; Nassauer and Opdam 2008 Mizgajski and Markuszewska 2010), even some kind of the "identity crisis" of landscape ecology, which according to Wu (2013) was perceived at the turn of the new millennium—force us repeatedly to reopen theoretical questions on definition of the landscape and its implementation in real policies.

Landscape is the object of endless interest, both professional and lay. It is an area into which we have to fit with all our interests, requirements, and ideas. We emphasize *with all*. The landscape is the only space we have available. It is a space that we use in multiple ways, a space in which we live, we reside, we work and we pursue our hobbies. It is the place from which we draw and burden with our requirements and the products of our activities.

We respect this space because of this and because of its manifold manifestations and forms. We admire it, enhance it and protect it. We find endless inspiration in the landscape.

Our diverse interests and requirements meet in the same space. Some interests are mutually compatible, other contradictory. Our activities and interests were previously handled more or less spontaneously; managed planning was gradually introduced and implemented. Planning, management, administration is based on the interests of urbanization, communication, agriculture, forestry, water management and recreation, along with (or against) the interests of conservation. As these planning activities occupy the same spaces, conflicts of interests, disputes and jurisdictional issues may occur. This has resulted in the need for an **integrated approach to landscape**, integrated care, management and planning.

One of the basic requirements of integrated landscape management is ecologically optimal spatial organization and use of the landscape (see below). The tools for applying it are specific **spatial planning processes**, which can therefore be considered as real tools of **integrated landscape management** (ILM). Their objective is the organization of human activities in the landscape as a whole; this holistic landscape space integrates all human activities.

The decisive step in this approach is the functional **integration of landscape-ecological knowledge** in a form of documentation into existing management and planning processes—particularly in physical planning, land use, agriculture, water and forestry planning and design (Haber 1990, Barsch et al. 1993; Zonneveld 1995; Cash et al. 2003; Kerényi 2007; Mizgajski and Markuszewska 2010; Nassauer 2012). With simplification we can state that we need to integrate two groups of knowledge: the first and initial is the definition what actually is to be planned and managed. This is the "demand" from humans. The second one is the knowledge of the **landscape**, the knowledge of what the landscape can offer, the landscape "supply". The confrontation of these two groups should lead to a process referred to as the **ecolo-**

gization of the landscape management, more specifically as the ecologization of spatial organization, utilization and protection of the landscape. The ecologization of spatial organization was one of the four key elements of the first environmental policies of Slovak Republic after the political changes in November 1989 (Miklós 1991). This process in Slovakia is also defined in practice as "landscape-ecological planning" as an obligatory part of physical planning. The process is supported also by Act 237/2000 Coll. amending and supplementing Act 50/1976 Coll. on territorial planning and the building code (the Building Act), under which the landscape-ecological planning is a "*comprehensive process of mutual harmonization of spatial requirements of economic and other human activities with the landscape-ecological conditions that result from the structure of the landscape. The ecologically optimal spatial arrangement and functional use of landscape, while ensuring satisfactory stability of the spatial landscape structure, protection and rational use of nature, biodiversity and natural resources, development and protection of the territorial system of ecological stability and the immediate human environment.* ***The structure of the landscape and its features*** *appear as limits, restrictions or supporting factors for desired activities in a given territory*" (§139a paragraph 4).

Ecologically optimal spatial organization based on an understanding of the **landscape as geosystem** (see below) must be based on reconciling the requirements of different or all sectors of human activity with the characteristics of all structures of the landscape, must consider the diversity of the conditions, as well as the capacity of the landscape for multifunctional use (Haber 2005; Brandt and Vejre 2004; Mander et al. 2005). A key aspect of the **geosystem approach** is therefore the description and determination of the role of the primary, secondary and tertiary landscape structure for management and spatial planning processes (see below). The geosystem approach to the landscape enables logical configuration of information and, consequently, its synthesis in the form of synthetic spatial units, spatial complexes—geosystems. These are the basic units for research as well as for the spatial planning processes.

For interventions in the landscape to be to the greatest degree possible in accordance with not only the present but also the primary structure of the landscape, for **change** of land use to cause the least negative consequences, these changes must be **rationally managed**. An essential element of all, even the most complex concepts is the **use of each individual area** and its **spatial arrangement**. The rational approach, which aims precisely to create the greatest harmonization of conditions for initial, secondary and human-desired structures of the landscape is the basis for the ecologically optimal organization and use of landscape.

It should be mentioned that the landscape as geosystem, its properties as regulatives as well as their role in Landscape Ecological Planning are also legally defined, namely in the Act 237/2000 Coll. The definition reads:

> Landscape is a complex system of space, location, geo-relief and other mutually functionally linked material natural, man-reshaped and man-created elements, in particular the geological substrate and soil-forming substrate, waters, soil, flora and fauna, man-made objects and elements of land use, as well as socio-economic phenomena in the landscape resulting from their connections. The landscape is the environment of humans and other living organisms. (§139a para. 3)

This legal definition of the landscape as a geosystem is based among others on long term scientific work of the authors of the presented book. Of course, this definition, the methods of the creation, assignation and demarcation of the geosystems are also broadly applicable in other countries.

1.2 Landscape and Its Integrated Management in Planning Processes

The landscape as a geosystem is a comprehensive integrated spatial-material entity—the environment for the life of people and other organisms rather than just its separate components. This principle was institutionally recognized at the highest level at the United Nations Conference on Environment and Development in 1992 in Rio de Janeiro, the Rio Summit 92 where Agenda 21 was created, namely in its Chapter 10 entitled “Integrated Approach to the Management of Land Resources”. This trend is clearly and simply defined: the landscape resources are used for different purposes for different activities that compete with each other and cause conflicts. But there is only one space which we must all fit into. This document justified the need as well as the explanation for the integrated approach to the landscape and defined the principles of **integrated landscape management** as a major instrument of sustainable development.

An integrated approach to landscape (as mentioned above) enshrined in Agenda 21 is called for by a number of papers, e.g. Preobrazhensky et al. (1980); Ružička and Miklós (1990); Caims et al. (1994); Richling et al. (1994); Miklós and Izakovičová (1997); Slocombe (1998); Szaro et al. (1998); Siebert (2004); Bezák (2006); Hreško et al. (2006); Huba and Ira (2006); Kozová (2006); Kerényi (2007); Izakovičová and Kozová (2008); Kertész (2010); Miklós et al. (2011a, b, c); Belaňová (2014); Belaňová et al. (2014); Miklós et al. (2015). However, it is a very complex issue. According to Antrop (2013) in this complex development, in the ‘policy landscape’ the real landscape is often the only integrating concept. In general, landscape research should become more applied, more society-oriented and less theoretical and academic.

In the conditions of the Slovak Republic (SR) the integrated approach to landscape was included in statute by Act 7/2010 Coll. on Flood Protection: §9 Coordination of the flood risk management plan with the River Basin Management Plan and other spatial planning tools, paragraph (1), which reads: “*In order to achieve optimal protection against floods draft measures, … flood risk management plan, as well as the river basin management plans must be coordinated with other instruments of planning for the area, particularly with land consolidation projects, land use plans, forest management plans so that they will together form a tool, of* ***integrated landscape management*** *for the entire area of the river basin.*”

The decisive aspect of the integrated approach is the mutual **harmonization** of two landscape subsystems - natural and socio-economic (Drdoš 1983; Izakovičová 2006).

This requires comprehensive research of landscape in its three basic dimensions (social, economic, environmental), exploring connections and relationships between different dimensions. It is essential to understand the landscape space as a complex resource and relationship within these resources that is able to satisfy human needs.

The integrated approach to landscape should help to achieve **comprehensive** management of the **entire area** and for **all activities** that should exist in that area. The explanation arises from the fundamental theses of landscape and an integrated approach to the use of it

- there is one landscape space, we have no other, therefore all activities must fit into the same space (state, county, region, municipality, cadaster);
- this space is filled with **material entities** of the geosphere (may be called differently, for example. complexes, geosystems, landscapes) and is thus a system of the integrated unity of space, location and all other interconnected physical components (elements) of geosystems;
- if an activity **damages one element** of the geosystem, **it disturbs the others**. This also causes a change to the overall functioning, ecological balance and stability of the landscape as well as changes to the potential of the landscape to fulfill the requirements and needs of society;
- at the same time there also applies the thesis: if we apply optimal measures for organization and land use, which are based on an integrated approach to protect the landscape as a whole, e.g., applying optimal localisation and management of green areas, meadows, pastures, we thereby **simultaneously** protect biodiversity, water quality, soil from erosion, accumulation of material, microclimate.

Under Chapter 10 of Agenda 21, as well as by elaboration of these principles, the skeleton and framework for all sectoral planning procedures should be a **physical (territorial, spatial) plan created on a landscape-ecological basis.** The landscape-ecological basis, along the lines of this requirement, is implemented in Slovakia into the process of physical (spatial) planning defined by Act 237/2000 Coll. amending and supplementing Act 50/1976 Coll. on territorial planning and the building code as the method of landscape-ecological planning (Landscape Ecological Planning—LANDEP (Ružička and Miklós 1982, 1990; Miklós and Špinerová 2011), as well as the methodology of Territorial Systems of Ecological Stability (TSES) (MoE 1993; Izakovičová et al. 2000; SkEA 2009). Without going to detailed explanation, the method of LANDEP is considering more the vertical-functional aspects of the landscape system, since the method of TSES also focuses on the horizontal relations.

All above-mentioned principles can be realized if the landscape is understood systematically,

landscape as a geosystem.

The LANDEP and TSES methods constitute the legally defined tools for **transmitting** landscape-ecological knowledge and information about geosystems in spatial planning processes.

References

AGENDA 21 (1992) United Nations conference on environment and development. Rio de Janeiro (United Nations), A/Conf. 151/4

Antrop M (2013) A brief history of landscape research. In: Howard P, Ian Thompson I, Waterton E (eds) The Routledge companion to landscape studies. Routledge, Oxon, New York, pp 12–22

Barsch H, Sauppe G et al (1993) Zur Integration landschaftsoekologischer und sociooekologischer Daten in gebietliche Planungen. Potsdamer Geographische Forschungen, Band 4. Universität Potsdam, p 226

Belaňová E (2014) Krajinno-ekologické aspekty integrovaného manažmentu krajiny v územnom pláne a v projekte pozemkových úprav. Dizertačná práca. Technická univerzita Zvolen, p 190

Belaňová E, Kanianska R, Kizeková M, Makovníková J, Jaďuďová J, Zelený J, Kočická E, Vaľková V, Wagner J, Mitterpach J, Samešová D (2014) Quo Vadis - čo a ako možno integrovať? In: Stav a trendy integrovaného manažmentu životného prostredia: vedecká monografia. In: Diviaková A (ed) et al. 2014, p 220

Bezák P (2006) Integrovaný prístup hodnotenia krajiny na príklade výskumu v NP Poloniny. In: Izakovičová Z (ed.) Smolenická výzva III. Integrovaný manažment krajiny - základný nástroj implementácie trvalo udržateľného rozvoja: zborník príspevkov z konferencie. Ústav krajinnej ekológie SAV Bratislava, pp 125–130

Brandt J, Vejre H (eds) (2004) Multifunctional landscapes: theory, values and history, vol I. WIT Press, Southampton

Caims J, Crawford TV, Salwasser H (eds) (1994) Implementing integrated environmental management. Virginia Polytechnic Institute and State University, Blacksburg, p 137

Cash DW et al (2003) Knowledge systems for sustainable development. Proc Nat Acad Sci 100(14):8086–8091

Chorley RJ, Kennedy BA (1971) Physical geography—a system approach. Prentice—Hall Interantional Inc, London, p 370

Drdoš J (ed) (1983) Landscape synthesis: geoecological foundations of the complex landscape management. VEDA, Bratislava

Haber W (1990) Using landscape ecology in planning and management. In: Zonneveld S., Forman RTT (eds) Changing landscapes on ecological perspective. Springer-Verlag, New York, pp 217–232

Haber W (2002) Kulturlandschaft zwischen Bild und Wirklichkeit. In: Schweizerisches Akademie der Geistes und Sozialwissenschaften. Bern, Akadenmievorträge, Heft IX, p 19

Haber W (2005) Pflege des Landes – Verantwortung für Landschaft und Heimat. In: Deutscher Rat für Landespflege (ed): Landschaft und Heimat. Schriftenreihe des Deutschen Rates für Landespflege, vol 77. Meckenheim, pp 100–107

Hreško J, Pucherová Z, Baláž I, Ambros M, Bezák P, Bírová L, Boltižiar M, Bridišová Z, Bugár G, Celuch M, David S, Gerhátová K, Jančová A, Kaločaiová M, Koštál J, Mederly P, Mišovičová R, Petluš P, Petrovič F, Rezník S, Rózová Z, Ružička M, Rybaničová J, Ševčík M, Trungelová D, Tuhárska K, Vanková V, Vereš J (2006) Krajina Nitry a jej okolia. Úvodná etapa výskumu. UKF Nitra, Fakulta prírodných vied, Nitra, p 182

Huba M, Ira V (2006) Integrované prístupy ku kraji ne a koncepcia trvalo udržateľného rozvoja. In: Izakovičová Z (ed.) Smolenická výzva III – Integrovaný manažment krajiny – základný nástroj implementácie trvalo udržateľného rozvoja. Zborník príspevkov z konferencie, Ústav krajinnej ekológie SAV, Bratislava, pp 89–93

Izakovičová Z (2006) Integrovaný prístup k hodnoteniu poľnohospodárskej krajiny. In Geografická revue: Geografické a geoekologické štúdie, 2006, roč. 2, č. 2, pp 333–339, ISSN 1336-7072

Izakovičová Z, Kozová M (2008) Integrovaný manažment krajiny - nástroj podporujúci udržateľný rozvoj územia. In Enviromagazín časopis o tvorbe a ochrane životného prostredia, roč. 13, mimoriadne číslo, pp 8–11, ISSN 1335-1877

Izakovičová Z, Hrnčiarová T, Králik J, Liška M, Miklós L, Moyzeová M, Pauditšová E, Ružičková H, Šíbl J, Tremboš P (2000) Metodické pokyny na vypracovanie projektov regionálnych územných

systémov ekologickej stability a miestnych územných systémov ekologickej stability. MŽP SR, Združenie KRAJINA 21, Bratislava, p 155
Kerényi A (2007) Tájvédelem. – Pedellus Tankönyvkiadó, Debrecen, p 184
Kertész Á (2010) Tájökológiai kutatások 2010. IV. Magyar Tájökológiai Konferencia, MTA FKI, Budapest, p 294
Kienast F, Wildi O, Ghosh S (eds) (2007) A changing world. Challenges for landscape research, vol 8, Landscape Series, Springer, Dordrecht, pp 753–754
Kozová M (2006) Strategické environmentálne hodnotenie ako nástroj udržateľného rozvoja regiónov. In: Acta Geographica Universitatis Comenianae, vol 47. pp 99–108
Krcho J (1968) Prírodná časť geosféry ako kybernetický systém a jeho vyjadrenia v mape. Bratislava, Geografický časopis, 20(2):115–130
Krcho J (1978) The spatial organisation of the physical-geographical sphere as a cybernetic system expressed by means of measure as entropy. In: Acta Facultatis Rerum Naturalium Universitas Comenianae, vol 16. Geographica, pp 57–147
Mander Ü, Müller F, Wrbka T (2005) Functional and structural landscape indicators: upscaling and downscaling problems. Ecol Ind 5(4):267–272
Miklós L (1991) Morphometric indices of the relief in the LANDEP methods and their interpretation. Ecology (ČSFR) 10:159–186
Miklós L, Izakovičová Z (1997) Krajina ako geosystém. VEDA, SAV, Bratislava, p 152
Miklós L, Špinerová A (2011) Krajinno-ekologické plánovanie LANDEP. VKÚ, Harmanec, p 159
Miklós L, Diviaková A, Izakovičová Z (2011a) Ekologické siete a územný systém ekologickej stability. Vydavateľstvo TU vo Zvolene, p 141
Miklós L, Ivanič B, Kočický D (2011b) Krajinnoekologická základňa integrovaného manažmentu povodia Ipľa. Digitálna databáza a tematické mapové vrstvy. Projekt HUSK 0801/2.1.2/0162 Vytvorenie jednotného monitoringu na báze priestorového informačného systému v povodí Ipľa. Banská Štiavnica, Esprit, spol. s r. o. (elektronický zdroj)
Miklós L, Izakovičová Z, Kanka R, Ivanič B, Kočický D, Špinerová A, David S, Piscová V, Štefunková D, Oszlányi J, Ábrahámová A (2011c) Geografický informačný systém povodia Ipľa: Katalóg GIS a výber máp. Bratislava: Ústav krajinnej ekológie SAV: Katedra UNESCO, Fakulta ekológie a environmentalistiky, Technická univerzita Zvolen, Esprit Banská Štiavnica, p 143
Miklós L, Kočická E, Kočický D, Diviaková A, (2015) Geosystémy ako krajinnoekologická základňa pre integrovaný manažment krajiny. (Geosystems as a landscape-ecological basis for integrated landscape management – in Slovak). Technická univerzita Zvolen. p 101 + Annexes
Mizgajski A, Markuszewska I (eds) (2010) Implementation of landscape ecological knowledge in practice. The Problems of Landscape Ecology. Volume XXVIII. Polish Association for Landscape Ecology, Wydawnictwo Naukowe Adam Miczkiewicz University, Poznań
MoE SR (1993) Metodické pokyny na vypracova nie dokumentov územného systému ekologickej stability. Bratislava. č. 101/93 – II
Nassauer JI (2012) Landscape as medium and method for synthesis in urban ecological design. Landscape Urban Plann 106:221–229
Nassauer JI, Opdam P (2008) Design in science: extending the landscape ecology paradigm. Landscape Ecol 23:633–644
Nassauer J, Santelmann MV, Scavia D (eds.)(2007) From the Corn Belt to the Gulf. Societal and environmental implications of alternative agricultural futures. Resources for the Future, Routledge, Washington p 272
Neef E (1967) Die theoretischen Grundlagen der Landschaftslehre. H. Haack, Gotha, Leipzig
Neef E, Richter H, Barsch H, Haase G (1973) Beitrage zur Klarung der Terminologie in der Landschaftsforschung Geographisches Institut der Akademie der Wissenschaften der DDR, Leipzig, p 28
Preobrazhensky V S, Minc A A (1973) Sootnoshenye ponyaty geosystema a ekosystema. In: Práce a materiály z biológie krajiny 20. Proceedings of 3rd International Symposium on the Landscape Ecological Research. ÚBK SAV, Bratislava

Preobrazhensky VS, Kupriyanova TP, Alexandrova TD (1980) Issledovanie landšaftnych sistem dľa celej ochrany prirody. In: Struktura, dinamika i rozvitije landšaftov, AN SSSR, Institut geografii, Moskva, pp 11–25
Richling A, Malinowska E, Lechnio J (eds) (1994) Landscape Research and its Applications in Environmental Management. IALE Polisch Assoc, Warsaw, p 289
Ružička M, Miklós L (1982) Landscape ecological planning (LANDEP) in the process of territorial planning. Ekologia (CSSR) 1(3):297–312
Ružička M, Miklós L (1990) Basic premises and methods in landscape ecological planning and optimization. In: Zonneveld IS, Forman RTT (eds) Changing landscapes: an ecological perspective. Springer—Verlag, New York, pp 233–260
Siebert SF (2004) Traditional agriculture and the conservation of biological diversity in Crete, Greece. Int J Agric Sustain 2:109–117
SkEA (2009) Metodické pokyny na vypracovanie aktualizovaných dokumentov R-ÚSES, CMŽP – OMK Bratislava, p 133
Slocombe DS (1998) Lessons from experience with ecosystem management. Landscape Urban Plann 40(1–3):31–39
Sochava V B (1977) Landscapes of Southeastern Siberia (1:1 500000 map). Moscow
Szaro CR, Sexton WT, Malone ChR (1998) The emergence of ecosystem management as a tool for meeting people´s needs and sustaining ecosystems. Landscape und Urban Plann 40:1–7
von Bertalanffy L (1968) General system theory. Foundations, development and applications. George Brazileer, Penguin Books, New York
Wu J (2013) Key concepts and research topics in landscape ecology revisited: 30 years after the Allerton Park workshop. Landscape Ecol 28:1–11
Wu J, Hobbs R (2002) Key issues and research priorities in landscape ecology: an idiosyncratic synthesis. Landscape Ecol 17:355–365
Zonneveld IS (1995) Land ecology: an introduction to landscape ecology as a base for land evaluation. Land management and conservation. SPB Academic Publishing, Amsterdam

Chapter 2
Landscape as a Geosystem

Abstract This is the core chapter of the book dealing with the theoretical principles of the geosystems. Defines the topical and choric models of geosystems, as well as the simplified model of the geocomplexes. There is explained the difference between state variables and typological characteristics of the elements of geosystems. Specific respect is given to the definition of the structures of the landscape. According to the genesis, physical character of the elements and according to the relation of structures to their role and management in planning processes we divide the landscape as geosystem to three substructures. **Primary landscape structure** is a set of material elements of the landscape and their relations that constitute the original and permanent foundation for other structures. These elements are mainly the elements of the abiotic sphere—the geological base and subsoils, soils, waters, georelief, air. **Secondary landscape structure** is constituted by human-**influenced, reshaped and created** material landscape elements that currently cover the Earth's surface. These are the elements of land use, real biota, man-made objects and constructions. **Tertiary (socio-economic) landscape structure** is a set intangible (non-material) socio-economic factors/phenomena displayed to the landscape space as interests, manifestations and consequences of the activities of individual sectors that are relevant to landscape. These are the protection and other functional zones of nature and natural resources protection, hygienic and safety zones of industrial and infrastructure objects, zones of declared zones of specific environmental measures, administrative boundaries, etc. Finally, the chapter gives the geosystem definition of the landscape and its reflection in the law in Slovakia. This definition reeds: "Landscape is a complex system of space, location, georelief and other mutually, functionally interconnected material natural elements and elements modified and created by a man, in particular the geological base and soil creating substratum, soil, water bodies, air, flora and fauna, artificial structures and the elements of land use, as well as their connections, which determine also the socio-economic factors related to landscape. Landscape is the environment of man and other living organisms." The chapter is illustrated by figures and graphics explaining the structure of the geosystem.

Keywords Geosystem · Geocomplex · Elements · Primary landscape structure
Secondary landscape structure · Tertiary landscape structure

L. Miklós et al., *Landscape as a Geosystem*,
https://doi.org/10.1007/978-3-319-94024-3_2

The concept of the landscape occurs in different sciences. Recently, slightly simplified, at least two main streams should be identified: the material entity/geosystem-based concepts of the landscape ("hard" concepts), and, the cultural-heritage, values, and perception-based ones ("soft" concepts). The first approach is represented mainly by geographers and landscape ecologists who grew up on geographical sciences, the second one by very different groups, which includes both specialists from landscape sciences, as well as very broad group of social scientists and even architects and artists. This book will concentrate on the **landscape as a geosystem** which may be regarded a **complex natural resource** for life and development of humans and other organisms. Its favourable vertical and spatial structure is a crucial aspect of the quality of the environment. Subsequently, the landscape as a geosystem should be a scientific base to the integrated landscape management which is the process of regulating the landscape use. This process requires integrated management tools that can absorb and properly use the landscape-ecological information on the geosystem.

2.1 The Approaches to the Definition of the Landscape

Of course, the different approach to the definition of the landscape is not a new issue. According to Naveh and Lieberman (1994) the landscape is historically perceived in two ways: as a tangible material reality and also as an intangible, mental and artistic experience, even as a way of the life (*genre de vive*, Vidal De la Blache 1922). A similar dichotomous understanding of the landscape has been expressed by many other authors as, e.g. Zonneveld (1981), Golley and Bellot (1991), Haber (2002, 2004, 2007), Hynek (2010). Authors as Grodzinski (2005), Hunziker et al. (2007) defined another dichotomy marked as space/place concept. However, for geographically educated landscape ecologists, the "space-places" word-pair evokes first of all the research dimensions—the choric and topic dimension (e.g. Haase 1973, 1980, 1996; Haase et al. 1991). These words evoke the same impression also in a common language (surely in Slavic languages) and for laymen. In addition, the first president of the International Association for Landscape Ecology IALE, Isaack Zonneveld, spoke about the huge diversity of landscape ecologists during the VIth International Symposium on Problems of Landscape Ecological Research (October 1982, Piešťany, Slovakia) where IALE was constituted. He considered landscape ecologists simply all those who deal with landscapes (*personal note of the author who attended to the Symposium*). Generally, there are permanently competing concepts such as geocomplex versus cultural landscapes, scientific versus cognitive approach, positivism versus constructivism (Bastian 2008; Antrop 2013).

The landscape has been the object of interest of landscape ecology since the works of Troll (1939), Bobeck and Schmithüsen (1949), Schmithüsen (1968, 1974). Landscape ecology has developed as a specialised integration of the disciplines of comprehensive landscape research (Chorley and Kennedy 1971; Mičian 1982; Preobrazhensky 1983; Risser et al. 1984; Forman and Godron 1986; Leser 1991, 1997; Finke 1994; Zonneveld 1995; Richling and Solon 1996; Nassauer 1997; Farina 1998;

Bastian 2001 Kertész 2002; Wu and Hobbs 2002; Oťaheľ 2004; Kerényi 2003; Haber 2004; Kienast et al. 2007; Kozová et al. 2007; Kolejka et al. 2011; Antrop 2013; Wu 2013 and many others). Nevertheless, the scientific conception of landscape as an object of research still has many different definitions, from understanding the landscape as an image up to a holistic understanding. Many scientific conferences and symposia have been devoted to clarifying the basic concepts of the landscape and landscape ecology, too. Let us mention the 3rd, 4th and 5th international symposia on the problems of landscape-ecological research organised in by the Institute of Landscape Ecology of Slovak Academy of Sciences (Neef et al. 1973; Proceedings 1973, 1976, 1979, 1982), up to the last one—the 17th in 1915 (Landscape …, 2012, 2015), or the congresses of the Czechoslovak geographers (e.g. their XVIth congress (Proceedings 1978), of the International Congress organised by the Netherlands Society of Landscape Ecology in Veldhoven (Tjallingii and De Veer 1982), and the Allerton Park workshop (Risser et al. 1984) held after foundation of IALE may be considered as constitutive ones. One acknowledgment of the scientific relevance of these conferences may be the fact that the International Association for Landscape Ecology was established at the 6th International Symposia on the Problems of Landscape Ecological Research in Piešťany (Slovakia), 1982. There are also a large number of newer scientific works, proceedings, recherché and compendiums analysing the concept of landscape (Grodzinski 2005; Longatti and Dalang 2007; Kertész 2010; Antrop 2013; Jones et al. 2013; Bruns et al. 2015). On other hand it has to be mentioned, that the significant diversification of the studies led to variable quality ranging from rigorous scientific analysis to almost pseudoscientific papers aimed at the broad public, sometimes applying innovations in amateurish way (Antrop 2013).

Deep analysis of this abundance of literature is not the intention of presented work. For the purpose of this book we confine ourselves only to a highly generalised formal division of possible approaches to the landscape. However, it has to be mentioned that the majority of recent recherché and compendiums paid much more attention to the landscape-ecological publications published in West-European and North-American countries than to those in Central or Eastern Europe (Csorba 1987). Our book tries to fill in this gap to a certain extent, too.

(a) Landscape as **Image**

The landscape as a landscape painting, an image of the area, a photo—as understood by the public and artistic sphere. In a slightly more specialised sense the landscape is a set of visual elements, especially relief, vegetation and other elements of land use, the scenery, spatial and aesthetic aspects of the landscape. Nevertheless, this approach also appears in many contemporary studies, even finding some support in the European Landscape Convention (2000).

(b) Landscape as a **Natural Complex**

The landscape in this sense is understood as a natural part of the geographic complex, the physical-geographical complex, without a socio-economic component. It is characterised by a range of physical-geographical features, from the geological base, up to vegetation and air. This understanding was characteristic, and still is popular, particularly in the Russian (Soviet) landscape-ecological school (the "landshaftovedenyje", Preobrazhensky and Minc 1973; Sochava 1977; Isachenko 1980, 1981; Preobrazhensky 1983; Snytko 1983; Grodzinski 2005; Puzachenko 2006; Snytko and Semenov 2008; Khoroshev et al. 2013).

(c) Landscape as a **Natural-Socio-economic Complex**

The landscape in this case is understood as a complex of physical-geographical and socio-economic components of the geographical sphere. This understanding is the basis for the application of theories of the landscape into practice, especially in spatial planning activities. The concept emphasises both the vertical and horizontal structure of the landscape units. This understanding was characteristic mainly for the German landscape-ecological school (Neef 1963, 1967; Neef et al. 1973; Schmithüsen 1976; Haber 1980; Snacken and Antrop 1983; Schreiber 1985; Haase 1996; Leser 1997; Brandt 1999). This approach has been widely applied and is also applied in the Central European landscape-ecological school (Demek 1974; Verrasztó 1979; Drdos et al. 1980; Mičian 1982; Csorba 1988; Richling and Solon 1996; Kistowski 1998; Oťaheľ 2004; Grodzinski 2005; Lowiczki and Mizgajski 2013; Veteikis et al. 2015). This approach later on developed into the understanding of the landscape as a geosystem (Krcho 1968; Demek 1974, 1978).

(d) Landscape as **Structure of Land Cover**

Landscape is in this case is considered to be the structure of the components of the current land cover and shape of the land emphasising the unity of the pattern and process, structure and function. This understanding of the landscape was characteristic mainly for the American landscape-ecological schools (Forman and Godron 1981; Risser et al. 1984; Turner 1990; Forman 1995). A number of quantitative method and spatial metrics were developed within this approach, as, e.g. entropy, heterogeneity, fragmentation of the landscape (Turner 1990; Turner and Gardner 1991; McGarigal 2002; Mezősi and Fejes 2004; Mander et al. 2005; Szabó et al. 2008; Csorba and Szabó 2012; Štefunková and Hanušin 2015). Within this approach emphasis on the horizontal relations of the land cover elements strongly prevails. This approach also became very popular in Western Europe and later all over the world (Schreiber 1980; Vejre and Brandt 2004; Wrbka et al. 2004; Veteikis et al. 2015). Some aspects of this approach are easy to understand for non-specialists too, e.g. a very popular subject of studies is the comparison, metrics and statistics of the changes of land use patterns over time. Leser (1991) also draws our attention to the horizontal and vertical aspects of the delineation of landscapes. According to him the horizontal aspect is trivial: for example, a micro-region as an ecosystem involving the typical processes, since the vertical aspect is less trivial, but at the same time extremely important. This approach also creates an important part of current international environment policy—e.g. the approaches to ecological networks (Jongman and Pungetti 2004; Brandt 1995; Csorba; 2008a; Schilleci et al. 2017). The approach is also characteristic for landscape architects (landscape engineering, landscaping). Nevertheless, the approach can fall to flattening of the scientific approach. Diverse landscape studies, sometimes with almost pseudoscientific quality, try to "offer 'innovation' in their domain, even when applied in a more amateurish way" (Antrop 2013).

(e) **Holistic Characteristics of the Landscape**

Many scientists have tried to express the holistic character of landscape, with greater or lesser success, often with difficult to translate phrases, e.g.:

- landscape as a region is the unity of human activity and the natural environment, the interaction of people and the environment in the region—***genre de vie***—genre of life (Vidal De La Blache 1922);
- landscape is a particular portion of the Earth's surface and is determined by the unit structure and similar character of the set of relations *(durch einheitliche Structur und gleiches Character Wirkungsgefüge geprägten konkreten Teil der Erdoberfläche* (Neef 1967);
- landscape is a particular space-time system of the total ecosphere (Naveh 1990).

In this direction there are a number of theories which are more or less philosophically tuned, without more precise definition of the subject (Teleki 1917; Szádeczky-Kardoss 1989), sometimes even understanding the landscape just as an aspect, a reflection of the real world (e.g. Zonneveld 1981; Golley and Bellot 1991; Hunziker et al. 2007; Hynek 2010). One can also include in this group the understanding of the total human ecosystem (Naveh and Lieberman 1984), the landscape as the home of the humans (Pedroli 2000), the "three-pronged" view of the landscape as "scenery–pattern–system" (Zonneveld 1995), or even the five-dimensional understanding of landscape as spatial–mental–temporal–natural/cultural–complex system understanding (Tress and Tress 2001). Nevertheless, however holistic the landscape is, for practical purposes it needs some concretisation (Verrasztó 2017; Antrop and Van Eetvelde 2017).

(f) Definition of Landscape from the **European Landscape Convention**

The classic example for a holistically constructed definition is the definition of landscape in the European Landscape Convention (Council of Europe, Florence, 20th October, 2000), which is of a rather declarative character and is not really practically suitable for use in management and planning, or, if you like, it allows a very liberal interpretation. It reads as follows (Article 1):

> Landscape means an area perceived by people, whose character is the result of the action and interaction of natural and/or human factors.

It is a non-materialistic definition; landscape is defined as an imaginary entity based on perception of its character. Other articles define the landscape as an assembly of "heritage", "values", "quality" (Article 5, 6). The problem is not the loose definition itself, but its acceptance, more precisely its non-acceptance in practice. Specifically, its acceptance is not practically controllable in "hard" policies, such as protection, management and planning, since whatever perception of the landscape of whomever might be considered as legal! Therefore the practice may apply the theoretical provisions of the Convention in a voluntary way (Antrop 2005). The Convention serves as the main pillar for the landscape ecologists who consider the landscape to be a phenomenon, the "scape" of the land, the cultural-heritage value. The specialists from this group do not always insist on the deep knowledge of landscape as geosystem, on the knowledge of the elements of landscape, of their physical structure (see, e.g. Breuste et al. 2009). To his approach can be ranked also the very popular approach of the evaluation and mapping of the "character" of landscapes, many times described as the mapping of landscape types (Wascher and Jongman 2000; Wascher 2005; Wrbka et al. 2005; Csorba 2008b; Wrbka 2009; Konkoly-Gyuró et al. 2010; Renetzeder et al. 2010), the mapping of "values" of landscapes and historical landscape structures (Špulerová et al. 2011; Štefunková et al. 2011).

Of course, differences in the understanding of specific studies and projects are not as clear-cut as we present here, and the geographical distribution of these approaches is not as sharp either (Hynek 2011; Žigrai 2015). There also is an apparent shift of "popularity" of different streams of understanding of the landscape in Central Europe, if looking at the content of the comparable, repeatedly held and traditional 17 International Symposia on Landscape Ecological Research organised by the Institute of Landscape Ecology of SAS (from Proceedings … 1973, 1976, 1979; up to Landscape … 2012, 2015), but also according to the content of other landscape-ecological and geographical symposia (e.g. Kozová et al. 2007; Breuste et al. 2009; Mizgajski and Markuszewska 2010; Kolejka et al. 2011; IX. Kárpát-medencei … 2013) or according to other sources (Longatti and Dalang 2007).

If further look deeper into various theories at their interpretation in practical trials, no matter how comprehensive and holistic, their view narrows to a much more simplified understanding. The narrow view is even more visible in application trials. In some cases, there is an apparent abandonment of even the simplified physical-geographical complexity and the result is analysis of only a few elements and relationships of the selected components of the landscape. Such an approach to the study of the landscape, although it may be scientific and the results may be very valuable, cannot be close to being considered a comprehensive or holistic approach to landscape.

2.2 The System Theory and the Landscape as a Geosystem

With the demands to understand the landscape holistically, considering all aspects mentioned above, we consider the most appropriate compromise between holistic theory and practical application of results, attempting maximum comprehensiveness of the study of landscape to be the understanding of landscape as geosystem, while still regarding the basic argument of the system theory that a system is more than a mere sum of its elements.

The geosystem theory is based on the general system theory that developed with gradual adaptation of the term "Gestalt" to geographical theory. We recall that term "Gestalt" is commonly accepted as one that cannot be translated exactly, but in any case it is a not exactly definable wholeness of landscape. On these principles, Austrian biologist and philosopher K.L. von Bertalanffy gradually developed his

General System Theory (von Bertalanffy 1950, 1968). His theory emphasises holism versus reductionism, organism versus mechanism, an open versus a closed system. Most simply, a system in his view is defined as

a system is a set of elements and their relations.

In landscape ecology and geography, the landscape is generally understood as a complex **material entity**, as a section from the geosphere, therefore, which comprises a set of ecosystems, under anthropocentric views as a space for human life. This character of the landscape is expressed by timeless definition of Neef (1967) that the landscape is a particular portion of the Earth's surface and is determined by the unit structure and similar character of the set of relations (*durch einheitliche Structur und gleiches Character Wirkungsgefüge geprägten konkreten Teil der Erdoberfläche*). In other words, it is characterised as a total system of the geographic sphere as a cybernetic system in the broad sense (e.g. Krcho 1968, 1974, 1991; Vološčuk 2003), concretised as a **geosystem** (Chorley and Kennedy 1971; Sochava 1978; Isachenko 1981; Haase 1973, 1980; Preobrazhensky and Minc 1973; Demek 1974, 1978; Mičian 1982; Neef et al. 1973; Preobrazhensky 1983; Snacken and Antrop 1983; Richling and Solon 1993; Naveh and Lieberman 1993; Miklós and Izakovičová 1997 and many others).

These basic postulates also clearly apply to the landscape as a geosystem that after the analysis of crucial aspects of definitions of a geosystem by various authors a simple, but comprehensive definition of a geosystem can be proposed:

a geosystem is a set of elements (components) of the geographical sphere and their mutual relations with each other

(Krcho 1968, 1978; Miklós and Izakovičová 1997; Miklós and Špinerová 2011). Of course, there are also other definitions of a geosystem that also use other system terms. E.g. often these definitions appear with concepts such as *structure, pattern of functioning, dynamics, matter, energy, information, synergy, spatiality, temporality.* All these terms are, however, implied as contained in the concept of **system**, or the terms **set, element relationship**.

This understanding corresponds to our understanding and definition of landscape (Miklós and Izakovičová 1997; Miklós and Špinerová 2011), which was also reflected in legal form in Act 50/1976 Coll. on territorial planning and the building code (Building Act), as amended by Act 237/2000 Coll.:

> Landscape is a complex system of space, location, landforms and other mutually functionally linked material of natural and man reshaped and formed elements, in particular the geological substrate and soil-forming substrate, waters, soil, flora and fauna, man-made objects and elements of land use, as well as connections resulting from socio-economic phenomena in the landscape. The landscape is the environment of mankind and other living organisms. (§139a paragraph 3)

Landscape as a geosystem may be regarded a **complex natural resource** for life and development of humans and other organisms. Its favourable vertical and spatial

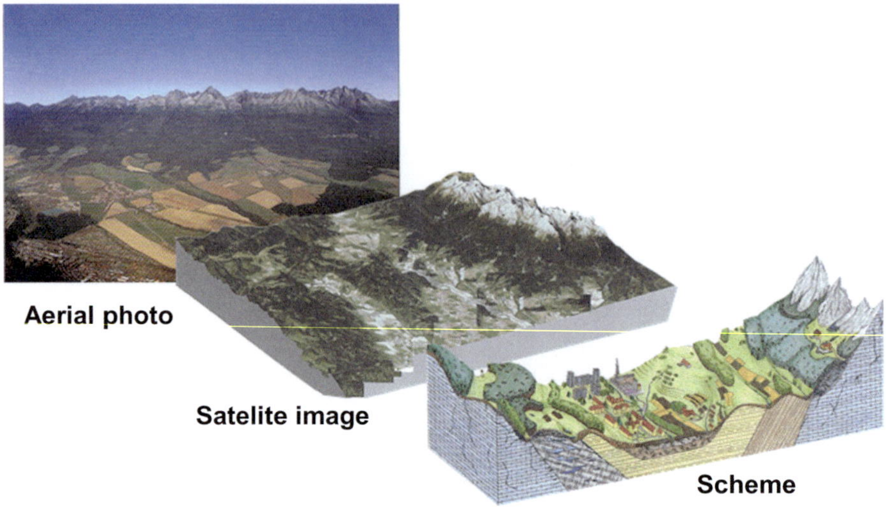

Fig. 2.1 Landscape as the material section from the geographical sphere and its models

structure is a crucial aspect of the quality of the environment. Subsequently, the landscape as a geosystem should be a scientific base to the integrated landscape management which is the process of regulating the landscape use. This process requires integrated management tools that can absorb and properly use the landscape-ecological information on the geosystem.

General theoretical and methodological aspects of the landscape as a geosystem, as mentioned above, are detailed in the works Miklós and Izakovičová (1997) and Miklós and Špinerová (2011). Further chapters deal with this understanding of the landscape, starting with methodical base of this approach up to presentation of a routine creation of database applicable for different landscape studies, spatial planning processes and **integrated landscape management** (Miklós et al. 2011a, b, c, 2014).

2.3 Models of Geosystems—Geosystems and Geocomplexes

A process for the optimal use of landscape at any point is predetermined by the synergistic effect of all the parametric property values of the landscape as a geosystem. Therefore it is extremely important to organise information about the geosystem in a suitable form and with an appropriate breakdown. The next items explain the essence of geosystems using models.

The models can be considered to be an abstraction of reality. In this respect, the simplest model of geosystems is their understanding as a material section from the geographical sphere (Fig. 2.1).

Applied approaches to landscape are characterised mainly by two kinds of models of landscape as a geosystem, the topic and the choric models.

2.3.1 Topical Model of a Geosystem

The elements of the model are the components of the geographical sphere of $\mathbf{a}_1$–$\mathbf{a}_n$. Relations in geosystems are labelled with the symbol $\boldsymbol{r}_n$. We can therefore write the model of the geosystem in the form

$$\mathrm{SG_K} = \{\mathrm{a}_n, \boldsymbol{r}_n\}$$

Such a topical model can be named also as the monosystem model (Preobrazhensky and Minc 1973; Preobrazhensky 1983). The topical geosystem model offers the simplest way to understand the vertical structure of the geosystems (Fig. 2.2), explains the **vertical structure** of the landscape as geosystem.

In the case of applied landscape-ecological works, the formal description of the relationship of each element with each other would be extremely difficult, not to mention that we could never know all the relationships. Therefore, for practical reasons, we approach work using

geocomplexes,

the material nature of which is of course identical with geosystems, but we write them formally only as a set of elements

$$\boldsymbol{G_K} = (\mathbf{a}_n),$$

whereby we implicitly attribute a set of relationships to the complex.

In this specific work **the topic model of geosystems is used**

- for **characterising elements of landscape** in the process of landscape-ecological analysis;
- **creation and characteristics of homogeneous geocomplexes** in the process of landscape-ecological synthesis (see later).

The Geosystem Approach to the Concept of Ecosystem

The most generally accepted characteristic of the ecosystem says that it is the system of living organisms and their surrounding elements (Tansley 1935; Odum 1975). This coincides with the principle of the scientific discipline of ecology, that addresses the relation of "dwelling" and "dweller" (oikos—house and inhabitant) and studies the relations of a central element—the dweller—most commonly a biotic component—to other elements of the "dwelling". It means that—according to the

Fig. 2.2 Topic model of a geosystem

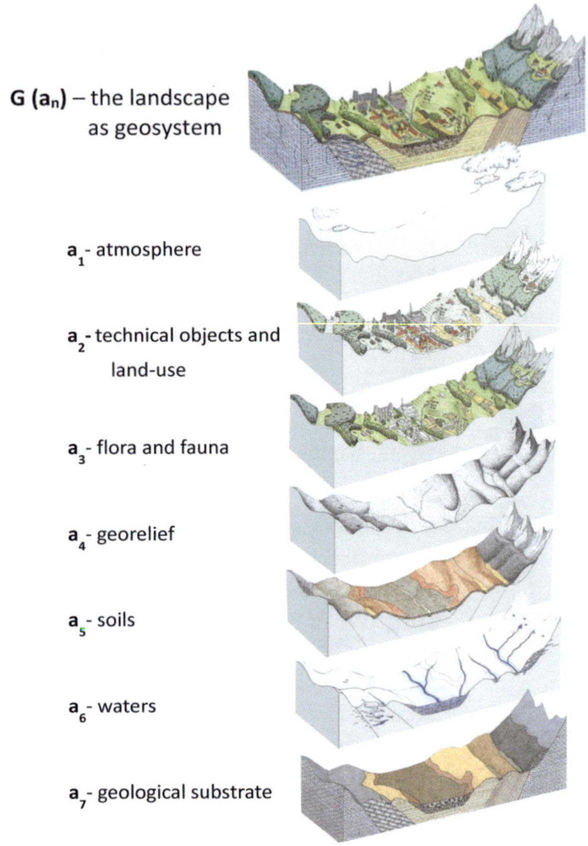

ecological theory—the living organisms are central in ecosystems. Accordingly, from the geosystem point of view, the model of the ecosystem can also be considered as a topical model and formally can be transcribed as follows:

$$\boldsymbol{E}_S = \{\mathbf{a}_n, \boldsymbol{r}_{3m-nm}\},$$

where the elements of the model are the same components of the geographical sphere $\mathbf{a}_1$ to $\mathbf{a}_n$, as in the geosystems, but in ecosystems the element $\mathbf{a}_3$—the biotic component (flora and fauna) is centralised and formally only relations $\boldsymbol{r}_{3m\text{–}nm}$ are assessed, which are the relations of all elements with the component $\mathbf{a}_3$ (Fig. 2.3). This definition of ecosystem is based on the understanding of the landscape as a geosystem where each material section of the earth's surface is the bearer of geosystems as well as ecosystems (Preobrazhensky and Minc 1973).

Of course, there are many other "classic" definitions of an ecosystem in biological disciplines, e.g. based on compartments (Ellenberg 1973; Odum 1975), but they do not affect the material essence of systems.

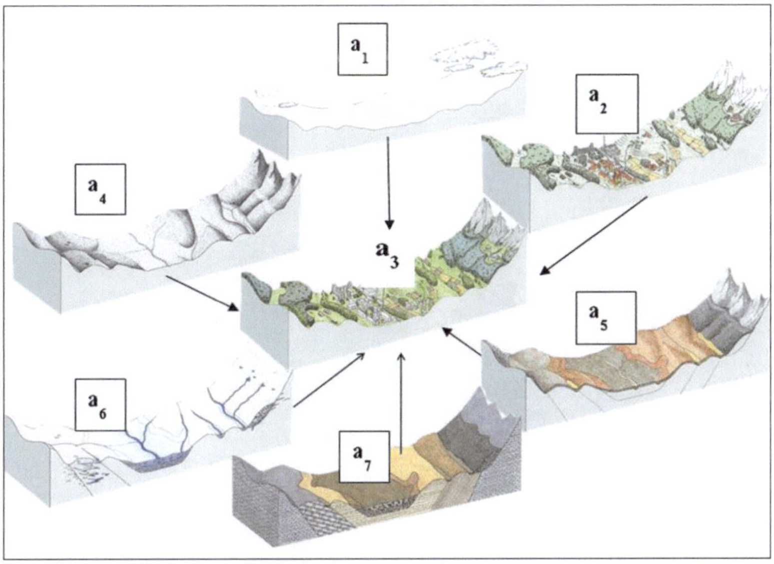

Fig. 2.3 Model of ecosystem from the geosystem perspective

It should be noted that, in practice, a clear-cut boundary between definitions of a geosystem and ecosystem approach is very rare and even less so between the geosystem and ecosystem approach in research. Most research is of an "ecosystem" approach character with a certain element centralised. This applies to mankind, where the most significant ecosystem approach stresses the protection of the environment—that of the centralised element of mankind and other living organisms This statement is also valid for studies on agroecosystems or forest ecosystems (Bunce et al. 1993; Mizgajski and Stępniewska 2012).

The Geosystem Approach to the Concept of Socio-economic Factors in the Landscape

In the landscape we can analyse and map a number of intangible entities and phenomena of the character of interests, manifestations and consequences of the activities of individual sectors that are relevant to landscape and have spatial manifestation. They display to the landscape the projection of the areas or boundaries of interest of all sectors as

- the nature conservation,
- protection of natural resources as waters, forests, soils,
- the interest areas of urbanisation and recreation,
- the zones of interest of agriculture, industry, transport, communal activities, including the areas of their negative impacts to the landscape in the form of hygienic and safety zones,
- the interests displayed in the form of boundaries of administrative or sectoral territorial units.

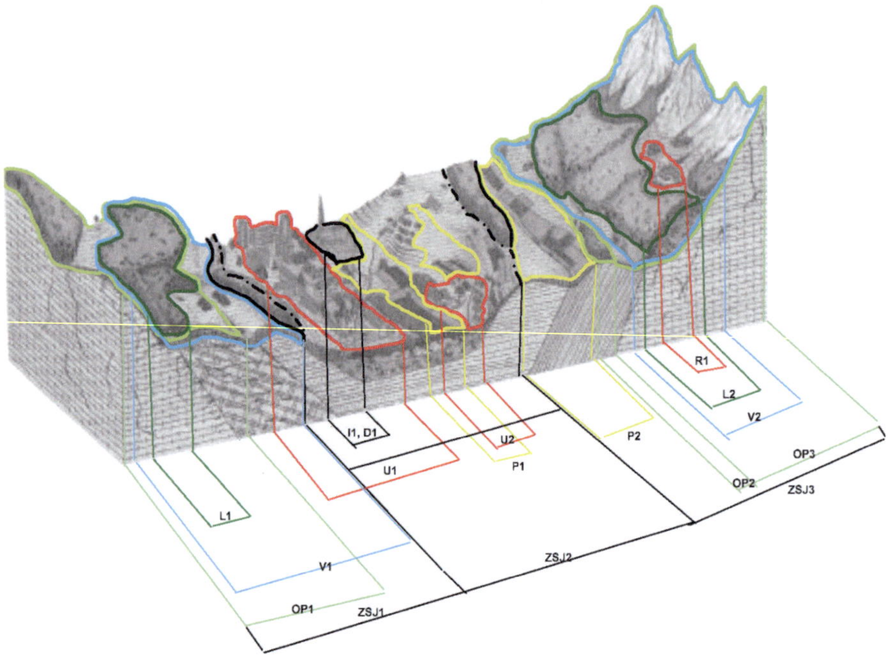

Fig. 2.4 Socio-economic factors in the landscape. SEF bounded to: I, D—industry and technical objects, U, R—urbanisation and recreation, V—protection of water resources, P—protection of high quality soils, L—forest resources protection, OP—nature conservation, ZSJ—administrative borders

The SEF themselves are intangible, not material but they are strictly bound to tangible elements of the primary and secondary landscape structure or their specific combinations.

For our purposes we label the concrete forms of the spatial manifestations of above-mentioned areas or boundaries of sectorial interests as socio-economic factors or phenomena in the landscape (**SEF**).

The model of the socio-economic factors is visualised on Fig. 2.4.

2.3.2 Choric Model of a Geosystem

The choric model divides the landscape to more or less homogenous parts according to defined rules creating spatial subsystems, so the elements of this model are the partial spatial subsystems (Krcho 1974, 1978) (Fig. 2.5), constituting the **horizontal/spatial structure** of the landscape as geosystem.

The model can also be named also as a polysystem model (Preobrazhensky and Minc 1973; Preobrazhensky 1983), because in addition to describing the system

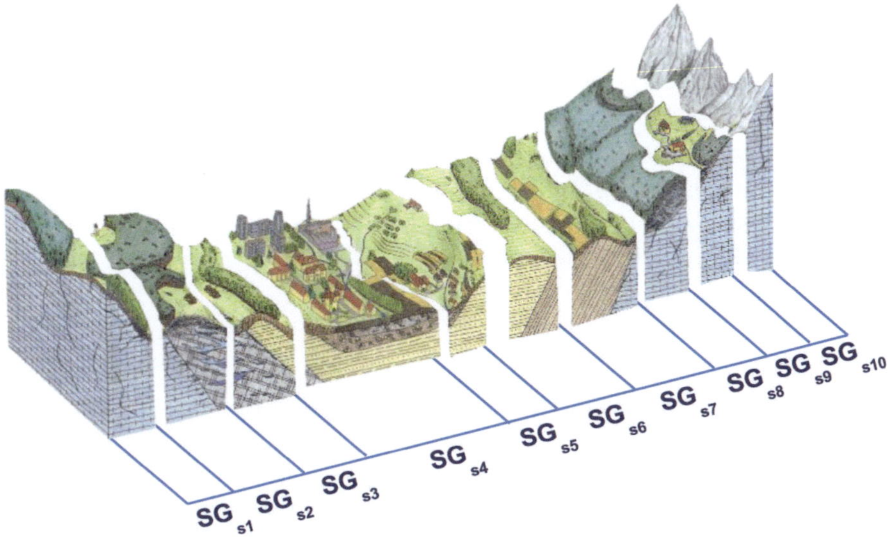

Fig. 2.5 Choric model of a geosystem

SG as a whole, each spatial subsystem $SG_{(n)}$ can also be described using the topic model G_K, as in the previous chapter. So, in this mode the polysystem model explains both the **vertical and horizontal/spatial structure** of the landscape as a geosystem. The choric geosystem model is also used often; it is the basis for the landscape-ecological syntheses, e.g. for the creation and characterization of abiotic complexes and landscape-ecological complexes in the method of the landscape-ecological planning LANDEP (Ružička and Miklós 1982; Špinerová 2015).

The recording of such a model has the form:

$$\mathbf{SG} = \{\mathbf{SG}_{(1)}, \mathbf{SG}_{(2)}, \ldots\ \mathbf{SG}_{(n)}\},$$

where elements of the model are partial spatial subsystems $SG_{(1)}$ to $SG_{(n)}$. A more precise expression of the multi-system form looks as follows:

$$\boldsymbol{S}\{\boldsymbol{G}_K(\mathbf{a}_n)\} = [\boldsymbol{S}\{\boldsymbol{G}_K(\mathbf{a}_n)\}_{(1)}, \boldsymbol{S}\{\boldsymbol{G}_K(\mathbf{a}_n)\}_{(2)}, \ldots\ \boldsymbol{S}\{\boldsymbol{G}_K(\mathbf{a}_n)\}_{(n)}]$$

In this specific work we use the characteristics of geosystems under the choric model using landscape-ecological synthesis, namely:

- in **landscape-ecological typification**; spatial subsystems $SG_{(n)}$ which have homogeneous content of elements $\boldsymbol{a}_n$. They can be understood as **types of geocomplexes** of topic character—abiotopes, biotopes, ecotopes (Mosiman 1984, 1990; Csorba 1988, 2014; Stanová et al. 2002; Diviaková 2011). The most commonly used are **abiotic complexes** (abiocomplexes, ABC), as the most stable part of geocomplexes (see below);

- in **landscape-ecological regionalization**: if spatial subsystems contain several different geocomplexes with different content, but are linked with defined bonds, it is possible to see them as regions of geocomplexes (**landscape-ecological regions**) (Neef 1963; Haase et al. 1991; Bailey 2002; Lowicki and Mizgajski 2013).

The conception of the topical and choric models of the geosystems respond to the "classic" approaches of topological and chorological research (Neef 1963*; Haase* 1973, 1980*; Haase et al.* 1991*; Bastian et al.* 2006*). Types and regions are widely used in current landscape-environmental work, as well as in spatial planning processes in Slovakia.*

2.4 Elements and Relationships in Geosystems

Elements of a **geosystem in topological or chorological units** are bound together by mutual relations in both vertical and horizontal directions (Schmithusen 1968, 1976; Mičian 1982; Szádeczky-Kardoss 1989; Khoroshev et al. 2013).

The physical substance of relationships in a geosystem is **energy-material and information flows** between elements of the geosystem.

The relations are displayed by state of the elements of the geosystems and their changes in time, which are caused by the processes and dynamics of the geosystems (Snytko 1983; Hrnčiarová and Miklós 1991; Hofierka and Šúri 1996).

On our distinguishing level, the elements of geosystems are **the components of the geographical sphere**. According to this approach, a geosystem consists of **elements**; we research these elements through indicators of their properties, which at each point in space have specific **values**. Strict differentiation of conceptual series

- system
– element of system
– property of element
– indicator of property
– value of indicator of property

is very important. Accordingly, the objective of landscape analyses is the creation and characteristics of indicators of properties of elements of landscape.

These values have the character of

- **state variables**—real measurable values of particular indicators of the properties of elements (e.g. the depth of soils, amount of rainfall, flow rate, tree height, amount of biomass), which determine the current status of an element;
- **typological characteristics** (spatial subsystems with typological values, e.g. soil type, climate type, physiognomic-ecological formation of plants). They are often expressed verbally and provide a comprehensive description of numerous properties of a particular element. These characteristics are commonly projected on maps in the form of spatial units;

- **combination of typological characteristics and state variables**.

It should be noted that it is impossible to lay down precise rules as to which elements should be characterised by state variables or typological characteristics. In general, it can be argued that state variables are suitable for large-scale landscape-ecological works as well as specialised studies, whereas typological characteristics seem to be preferred in informative and descriptive studies and in less detailed works on smaller scales.

The relations, the energy-material and information flows in geosystems, which can be also called **processes**, can be determined by

- **measuring** the values of the state variables of indicators of properties of those elements of the geosystem which affect the examined relationship—for example, measuring the amount of rainfall and the amount of soil washed away in determining the relationship of precipitation and soil erosion. This method of determining relationships is typical for specialised analytical geography, environmental science and other disciplines, in which it is the assessment of the relationship between selected elements that is the main subject of research. In this way, the individual—specialist can study a few relationships in-depth while trying to determine as closely as possible the values of material and energy flows.
- **comparison** of the values of state variables of one element to the value of the state variables of another element—e.g. determining (measuring) the altitude and determining the plant community to establish a relationship between altitude and vegetation. In this case we do not search the real cause of relationship, exploring the nature of energy and material flows, which are obviously very complex. We are content with the fact that we know the results of these relationships based on years of specialised analytical studies, subsequent comparison of the characteristics of a synthetic evaluation of geosystem elements (Tarboton 1997; Guth and Kučera 1997; Špinerová 2015). Such knowledge is also characteristic for landscape ecology, which often works well with "soft" systems with data sets that are referred to as "fuzzy data sets", which recognises that in the evaluation of relations geosystems we work—in relation to the level of perfection of its knowledge—also with a "grey" or "black" box.

The ecological sciences, including landscape ecology also often use the term "**autoregulatory mechanisms**". Essentially, autoregulatory mechanisms govern energy-material-information flows, which, in space and time, maintain certain conditions in geosystems. Frequently they are understood as positive processes which occur mostly in natural systems. However, it needs to be emphasised that autoregulatory mechanisms are constantly at work, in any conditions, in primeval forests, deserts, heaps, sewers, regulated watercourses and even tarns. Man can change some indicator values of the properties of the elements, which can disrupt, accelerate or hamper the process of autoregulatory mechanism, but he cannot eradicate them. For example, man can regulate CO_2 emissions but he is not able to prevent the greenhouse effect of the atmosphere, can change the soil surface by heaping a tailings pile, but he cannot prevent the growth of pioneer plants on the pile. Even if he tried to exterminate them with another layer of tailing rock and thereby leave the surface exposed, he cannot prevent erosion from happening on that surface. There are countless examples we could mention, but the point here is to provide a reminder that the process is an important aspect of the geosystem.

2.5 Structure of Landscape as a Geosystem

For research, as well as for practical purposes, a very important aspect of the geosystem approach to landscape is the characteristic of the **landscape structure**. There is a number of works devoted to this issue where one can find different approaches to the understanding of the structure. The most popular—probably also the easiest understandable—approach is the characteristic of the spatial structure of the land use, the characterisation of the pattern (Forman 1995; Turner 1990). This approach led to development of an amount of quantitative methods and metrics within landscape ecology (Turner and Gardner 1991; Gustafson 1998; Mcgarigal 2002; Oťaheľ et al. 2004; Mezősi and Fejes 2004; Csorba and Szabó 2012). According to this approach the landscape structure is the inherent configuration of the quantitative and the qualitative phenomena of landscape, reflected in complex physiognomic-functional clusters (Szabó 2007, 2008; Šteffek et al. 2008; Špulerová et al. 2011).

The other approach is close to the physical-geographical principles and also emphasises the vertical functional structure of the landscape as geosystem (Neef 1967; Krcho 1968; Isachenko 1981; Mičian 1982; Mosimann 1990; Haase et al. 1991; Bastian and Schreiber 1994; Snytko and Semenov 2008; Csorba 2014; Christopherson et al. 2016).

The complex—vertical/horizontal—landscape structure should be used as for basis for classifying the landscape as a geosystem, whether by typification or regionalization (Ružička et al. 1978; Ružička 2000; Bailey 2002; Csorba 2008b; Kolejka et al. 2011; Lowiczki and Mizgajski 2013; Štefunková and Hanušin 2015).

According to the genesis, physical character of the elements of the geosystems (see above) and last but not least according to the relation of structures to their use and management we divide the landscape as geosystem according to the topical model of the geosystem (see above) into three substructures (Miklós and Izakovičová 1997; Miklós et al. 2011a, b, c; Špinerová 2015) (Fig. 2.6).

2.5.1 *Primary Landscape Structure*

Primary landscape structure (PLS) is a set of material elements of the landscape and their relations that chronologically constitute the original and permanent foundation for other structures. Notable characteristic PLS elements are

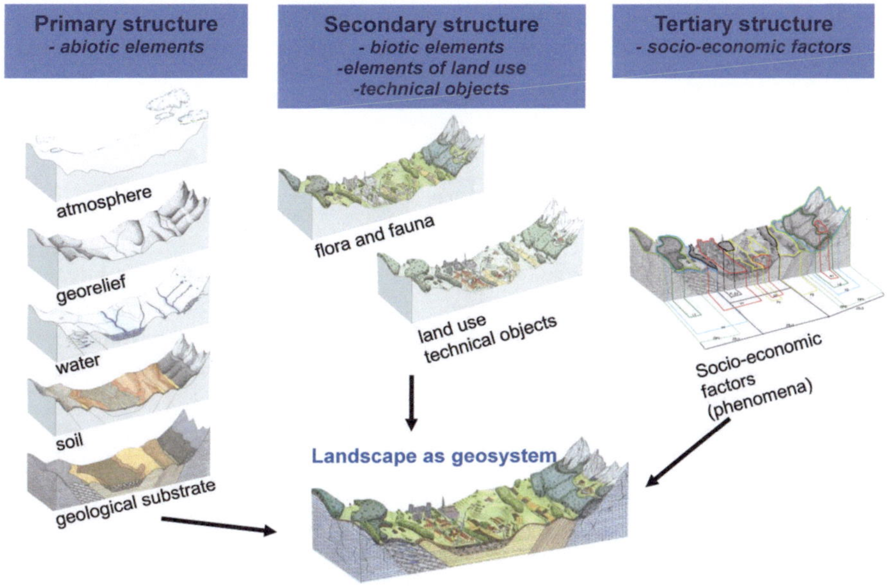

Fig. 2.6 Structure of the landscape as a geosystem

- physically **bound** to a particular place on Earth;
- **the principles of their functioning are unchanged by mankind** (e.g. water flows in the direction of gravity, south-facing slopes get more heat than north-facing slopes, the prevailing wind direction is set, the weathering of rocks cannot be stopped, etc.), but on the other hand;
- their reaction after disturbance is hard to control;
- **the material and structural basis is very hard or impossible to change** (e.g. limestone is not granite, where mountain ranges are, there are not lowlands, cold climate is not warm climate, etc.);
- these elements have in practice been **least changed** by mankind in comparison with the secondary and tertiary structure of the landscape, which mankind has **directly created**.

This structure is formed by the abiotic elements of the geosystem: geological substrate, subsoil, relief, waters, air. **(Geo)relief** has a specific character: it creates the phasal interface between the gaseous, liquid and solid phase of this structure itself, is intangible, represents the surface forms (Krcho 1968, 1974, 1991). Examples see in next corresponding chapters and tables (Tables 3.1a, b, Sect. 3.1.1).

A specific interpretation of the primary structure (of the abiotic complex) based also on the knowledge of real vegetation is the concept of potential natural vegetation. This means the "potential" vegetation does not really exist, where any vegetation does exist it is already real vegetation that has occurred secondarily (see examples in the following chapters, Table 3.1b). Complexes (communities) of original natural vegetation and fauna in our conditions are virtually absent. Even communities with quasi-natural substances are to some extent affected by humans, but most are commonly changed.

2.5.2 Secondary Landscape Structure

Secondary landscape structure (SLS) is a set of physical elements that have been formed secondarily, by human activities reshaping the primary landscape structure. It consists of a set of human-**influenced, reshaped and mankind-created** material landscape elements that currently covers the Earth's surface (Ružičková and Ružička 1973; Miklós and Izakovičová 1997). As for influenced elements we rank, e.g. the forests, as reshaped mainly as agricultural land, and as newly created the buildings and other technical objects.

Major characteristics of the elements of the SLS

- they are **physiognomic** elements of the land cover;
- they are physically **bound** to a specific place on the surface of the earth;
- they are **changeable** with a certain amount of energy (e.g. forest to fields, fields to buildings, buildings to scrub, etc.).

Elements of the SLS can be characterised from the perspectives of

- Land use forms (physical forms of **land use**) or **land cover**—physiognomic-functional perspective;
- their biotic content (**real vegetation and fauna**)—physiognomic-ecological perspective;
- their **spatial structure**—structural-spatial perspective (Miklós and Špinerová 2011).

The SLS is bound and dependent on the components of the primary landscape structure (Ružičková and Ružička 1973). At the same time it has firm relations to the tertiary landscape structure.

Between the terms as secondary landscape structure, land use and land cover there is a causal relationship. All forms and manifestations of the secondary landscape structure—from the so-called cultural landscape, which landscape-ecological studies consider an ideal state, to built-up industrial areas—came to existence by means of **land use** activities. Therefore, the secondary structure is the spatial manifestation of land use activities. For landscape-ecological evaluation it is important to consider the term **current landscape structure (CLS)**, which is understood to means the secondary landscape structure at present. The current landscape structure has been created as a result of land use, therefore, according to economic and geographical terminology, its elements are also forms of land use (Ružička et al. 1978; Žigrai 1983, 1995). Elements of the current landscape structure in various works are referred to as elements of land cover. The land cover is seen as the "visible" layer of landscape sphere, as the physiognomy of the landscape (Feranec and Oťahel 2001). At the same time, these elements can be characterised with varying degrees of detail and according to their bio-organic content as physiognomic-ecological formations of real vegetation as well as biotopes or habitats. In planning and other operating procedures they are referred to simply as mapping units of CLS. The basic classification of mapping units of CLS still follows the original division of the secondary landscape structure, as proposed by Ružička and Ružičková (1973). The only difference is the level of detail which is determined by the objectives of a specific project (examples see in next chapters, Table 3.2). The secondary landscape structure is where mankind has the most direct interest in the result of changes (Drdoš et al. 1995). Therefore, any planning of the optimal ecological organisation and use of land as well as protection of nature and natural resources is possible only by establishing a manner of use of land for every individual area of the territory (Haber 2005, 2007; Štefunková et al. 2011; Špulerová et al. 2011). So, we can state that the elements of SLS/CLS are the main operational units for planning and management procedures.

Examples see in next corresponding chapters and tables (Tables 3.2 and 3.4a–e, Sect. 3.1.1).

2.5.3 Tertiary Landscape Structure

Tertiary (socio-economic) landscape structure (**TLS**) is a set socio-economic factors/phenomena (SEF) displayed to the landscape space. As described above SEF are intangible entities and phenomena of the character of interests, manifestations and consequences of the activities of individual sectors that are relevant to landscape and have spatial manifestation. So, the **SEF** can be named as elements of TLS.

Important characteristics of the tertiary landscape structure are

- the SEF intrinsically (by themselves) are **intangible**, but **bound to tangible elements** of primary and secondary landscape structure;
- must be of **landscape-ecological relevance**, i.e. they have **spatial expression** (they are "mappable" in the space).

As SEF are **intangible**, there is no sense in speaking about their **physical changeability** (simply they do not physically exist). In spite of this, they very **significantly influence** the utilisation of the landscape use, present and future, since they are very closely bound to human interests.

The elements of TLS can be labelled with the term **socio-economic factors**/phenomena in the landscape (**SEF**). They are defined in the regulations with differing legal force—the acts, regulations, directives, standards, codes of practice, conventions, and development documents such as plans, projects, programmes of economic and social development of municipalities, local Agenda 21, documents of territorial systems of ecological stability, governmental development concepts and so on. SEF has spatial manifestation in the character of zones, sections, bands, sites, regions, protected areas, which are defined by law or other documents of varying force. They apply to those areas of human activities that have spatial demands. SEF are carriers of the **guidelines, restrictions and prohibitions** on human activities (Miklós and Špinerová 2011). On the basic level we can define the groups of SEF according to their character as

- boundaries and territories of declared nature conservation areas;
- boundaries and territories of declared protected areas of natural resources, zones of hygienic protection of water resources;
- protection zones and the safety zones of production, transportation and other technical facilities;
- administrative boundaries and sectorial boundaries;
- boundaries of sites, sections or territories of declared deterioration of the environment.

More detailed characteristics and examples see in next corresponding chapters and tables (Tables 3.3 and 3.4a–e, Sect. 3.1.1).

2.6 Interrelationships of the Individual Landscape Structures

An important aspect of the division of the landscape into the above-mentioned structures is their mutual relation, especially with respect to **planning and projecting practice**.

The decisive impact of these three structures on planning of activities (management) in the landscape, according to the logic of their characteristics should be as follows:

- as the primary landscape structure PLS has immutable principles of operation, impossible and difficult to change properties, but easily changeable quality, **planning should primarily adapt utilisation of landscape to its characteristics, where possible not to change them**;
- the secondary/current landscape structure SLS/CLS is changeable by using an amount of energy. Planning can therefore **consider its changes**, but with respect to its quality, if possible according to the properties of the primary landscape structure;
- as the tertiary structure does not physically exist, in theory it is the easiest to change. Therefore it should be carefully **adapted to the characteristics of primary and secondary landscape**.

Briefly: from the point of view of environmental care, nature conservation, maintaining ecologically optimal management and utilisation of land, as well as from the point of view of planning processes, **unchangeable and partially changeable landscape, the PLS and SLS/CLS**—the current landscape-ecological conditions—are the most critical structures because their disruption causes all ecological problems, whereas tertiary structure of the landscape TLS—the requirements of humans—**should adjust** to the primary and secondary landscape structure.

However, the last-mentioned statement is a difficult issue. Naturally, the occurrence of the elements of the tertiary landscape structure is not incidental. As mentioned above, socio-economic factors and SEF phenomena, although intangible, are linked to specific tangible elements, or specific combination of material elements of the primary or secondary landscape structure. Some SEP always co-occur with a certain element of the primary or secondary landscape structure; some occur only in specific territories. Some examples: In case of the occurrence of a protected plant or animal species, their territory is linked to a SEF—protected habitat. On the other hand, not every territory which exhibits a precious landscape structure similar to already declared national parks will automatically be declared as protected landscape area or a national park. Not every territory with rich water resources will be declared a protected water management area—e.g. the entire Danubian Plain is extremely rich in groundwater, but not the whole territory was declared as a Protected Water Management Area. On other side, for example, each transport line, product line or energy duct has its own protections zone designated, and each farm has a sanitary zone and every quarry has its safety zone, etc. Given the large amount of SEF and their possible combinations in the terrain, it is not possible neither feasible to provide a comprehensive overview of this situation, and it is possible only in a particular territory.

In practice, these principles are reflected in a different order

- if something is defined by the tertiary structure (socio-economic phenomena), for example acts, regulations, plans, regardless of the correctness of their existence, for planning they are many times almost unchangeable principles, because the authorities insist on these principles very strongly;
- the secondary structure is variable, so if it is not limited by any element of tertiary structure, it is only a slight obstacle to planning;
- in large areas of territory, the primary structure is not limited by any regulations, e.g. relief, climate, etc. so planning only respects it if it conflicts with tertiary or secondary structures.

With a certain degree of simplification, it can be claimed that the SEF from the group of transportation, communal-technical activities, industrial and mining activities and agricultural activities are usually linked to protection, sanitary or buffer zones. SEF from the group of nature conservation as well as recreation and housing are not necessarily always linked to any elements of the primary or secondary landscape structure. SEF of administrative territorial division are also not permanently linked to certain elements of the primary and secondary structure. SEF of deterioration of natural resources and the environment are linked only to an incidence of heightened concentration of pollution or other deterioration.

2.7 Definition of the Landscape and Its Reflection in the Law

Based on the definitions proposed by various authors, a landscape as a geosystem can be defined as a complex of components (elements) of the geographical sphere and their interrelationships. From the **systemic and factual perspective**, this definition can refer to the same material entities as, e.g. the terms landscape, *geoecosystem, geocomplex, geographical complex*, but also other terms such as *the territory, region, river basin, environment,* etc. whereas it is still the same particular material section of the geographical sphere—but with different borders. By other words we can recognise on each particular material section of the geographical sphere at the same time geosystem, geocomplex, territory, region, river basin, etc.

The landscape as a geosystem for our practical purposes is defined as follows:

Landscape is a complex system of

- ***space, location, georelief***
  and other mutually, functionally interconnected
- ***material natural elements and elements influenced, modified and created by man,***
  in particular geological base and soil creating substratum, water bodies, soil, flora and fauna, artificial objects and the elements of land-use, as well
  as by their relations determined
- ***socio-economic factors (phenomena) in the landscape.***

The landscape represents the environment of man and other living organisms.”
(Miklós and Izakovičová 1997; Landscape Atlas of Slovak Republic (Miklós et al. 2002), Act No. 237/2000 Coll., §139).

The above wording of the definition of the landscape was also adopted in Act 50/1976 Coll. on Territorial Planning and Building Order as amended by Act 237/2000 Coll., in §139. Moreover, the same Act defines also the elements and structure of the landscape as regulative of spatial arrangement and functional use of land, as:

§139 Terms of land-use planning

(1) … a regulative of spatial arrangement and functional use of land is a binding directive, which guides the location and arrangement of a certain object or implementation of a certain activity in a territory. It is expressed by the ***values of properties of elements of landscape structure*** *in words, numbers and, if applicable, also graphically. The regulative has the character of bans, restrictions or supporting factors in relation to spatial arrangement and functional use of the land. …*

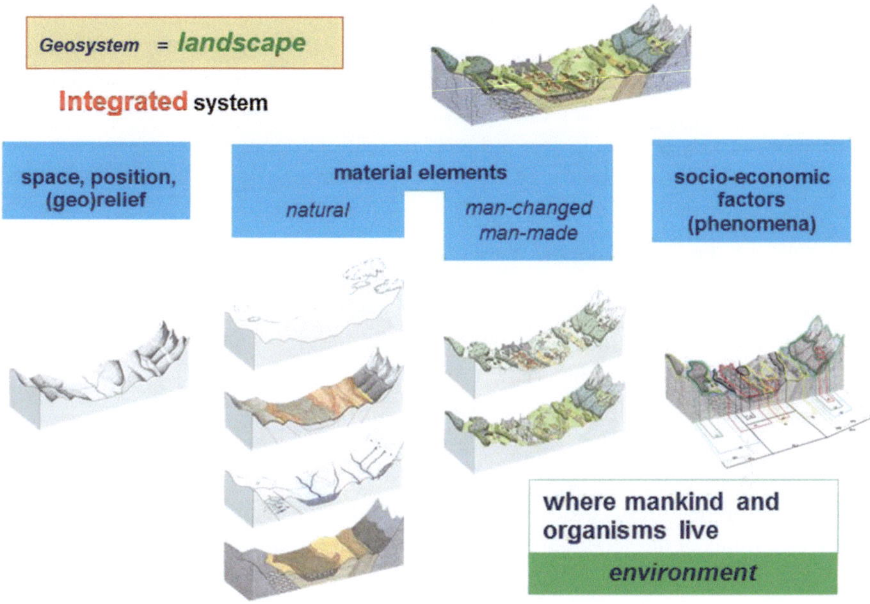

Fig. 2.7 Definition of landscape as a geosystem

The definition can be graphically represented as follows (Fig. 2.7).

So, if the same object is also seen in the subject-object relation, the landscape is considered the environment of life and activity of man, and thus the definition of the environment is the same: *the landscape as a geosystem is the environment of human life and activities.* Thus, the environment is not discrete, but the same object, the landscape space filled with material components of the landscape where humans and other living organisms exist.

Other utilitarian perspectives can interpret landscape structures also as

- life-sustaining conditions on earth;
- material source and potential for life of man;

but also as

- the setting for "the conflict of interests" of various activities;
- a critical aspect of environmental care.

The landscape structure can be considered a complex natural resource enabling life of humans and other organisms. For practical reasons it is appropriate to distinguish between material landscape resources and landscape potentials as follows:

Material resources: include the elements of the geosystem, providing material and energy for reproduction processes They are objectively existing, their properties can be measured and, if necessary, utilised.

Landscape potentials (of suitability for): prerequisites of utilisation for various purposes. The potential of the landscape as a whole is defined as the ability of the landscape structure to meet the requirements and needs of the society in the event society decides to exploit such ability (Haase 1978; Miklós 1988; Tremboš 1994). So, they occur only if there is demand (e.g. recreational potential).

According to various landscape structures, the most important resources and potentials can be outlined as follows:

(a) Primary landscape structure

- material resources: air, water, minerals, soil;
- potential: gravitational and positional potential: relief, space, location—as unalterable and unenlargeable resources;
- bioenergetic potential, heat-moisture regimen.

(b) Secondary landscape structure

- material resources: biotic resources, technical structures;
- potentials of the secondary landscape structure as a whole to provide qualities for the life (e.g. recreational potential);
- reproductive potential of landscape elements, carrying capacity, stability, resistance, resilience, etc.

(c) Tertiary landscape structure

- human resources;
- socio-economic potential: to satisfy the interests of production industries, nature conservation and protection of natural resources.

More recently, these same aspects of the landscape—material goods, suitability for different forms of utilisation, potentials, socio-economic and environmental functions of landscape have been referred to as ecosystem services (Costanza et al. 1997; Kienastet al. 2007; De Groot et al. 2010; Iverson et al. 2014; Grunewald and Bastian 2015).

References

Antrop M (2005) From holistic landscape synthesis to transdisciplinary landscape management. In: Tress B, Tress G, Fry G, Opdam P (eds) From landscape research to landscape planning: aspects of integration, education and application, UR Frontis Series, vol 12. Springer Science+Business Media, Wageningen, 5 Sept 2011, pp 27–50. http://library.wur.nl/frontis/landscape_research/03_antrop.pdf.accessed

Antrop M (2013) A brief history of landscape research. In: Howard P, Ian Thompson I, Waterton E (eds) The Routledge companion to landscape studies. Routledge, Oxon, New York, pp 12–22

Antrop M, Van Eetvelde V (2017) Landscape perspectives. The holistic nature of landscape. Springer, Netherlands, p 372

Bailey RG (2002) Ecoregion-based design for sustainability. Nova Iorque: Springer-Verlag, New York
Bastian O (2001) Landscape ecology—towards an unified discipline? Landscape Ecol 16:757–766
Bastian O (2008) Landscape classification between fact and fiction. In: Lechnio J, Kulczyk S, Malinowka E, Szumacher I (eds) Landscape classification. Theory and practice, Warshaw, pp 13–20
Bastian O, Schreiber KF (1994) Analyse und ökologische Bewertung der Landschaft. Gustav Fischer Verlag Jena - Stuttgart, p 502
Bastian O, Kronert R, Lipsky Z (2006) Landscape diagnosis on different space and time scales—a challenge for landscape planning. Landscape Ecol 21:359–374
Bobeck H, Schmithüsen J (1949) Die Landschaft im logischen System der Geographie. Erdkunde 3:112–120
Brandt J (1995) Ecological networks in Denmark. In: Van Buuren M, Jongman R (eds) Ecological networks in Europe, Special issue of the journal "Landschap", Wageningen, pp 63–76
Brandt J (1999) Geography as 'landscape ecology' Dan J Geogr 1(Special Issue):21–32
Breuste J, Kozová M, Finka M (eds) (2009) European landscapes in transformation: challenges for landscape ecology and management european IALE conference 2009, Salzburg (Austria), Bratislava (Slovakia)
Bruns D, Kühne O, Schönwald A, Theile S (eds) (2015) Landscape culture—culturing landscapes. The differentiated construction of landscapes. http://www.springer.com/de/book/9783658042837
Bunce RGH, Ryszkowski L, Paoletti MG (1993) Landscape ecology and agroecosystems. Lewis Publishers Conference publication, Boca Raton, FL, p 241
Chorley RJ, Kennedy BA (1971) Physical geography—a system approach. Prentice—Hall Interantional Inc, London, p 370
Christopherson RW, Cunha S, Thomsen CE (2016) Geosystems core. Pearson, p 512
Costanza R, d'Arge R, de Groot R, Farber S, Grasso M, Hannon B, Limburg K, Naeem S, O'Neill RV, Paruelo J, Raskin RG, Sutton P, van den Belt M (1997) The value of the world's ecosystem services and natural capital. Nature 387:253–260
Csorba P (1987) A tájökológia időszerű kérdései az angol nyelvű szakirodalom alapján. Földrajzi Közlemények 35 (111) (1–2):74–80
Csorba P (1988) The ecogeographical pattern map of Tokaj "Great Hill" and its environs. In: Ruzicka M, Hrnciarová T, Miklós L (eds) Spatial and functional relationships in landscape ecology. VIIIth international symposium on problems of landscape ecological research. IEBE CBES of SAS. IALE East European Region, vol 1. Bratislava, pp 273–284
Csorba P (2008a) Landscape ecological fragmentation of the small landscape units (Microregions) of Hungary based on the settlement network and traffic infrastructure. Ekológia (Bratislava) 27(1):99–116
Csorba P (2008b) A tájhatárok kijelölése és változása. Földrajzi Közlemények 132(2):220–226
Csorba P(2014) A tájszerkezet földrajzi értelmezése. A geographical interpretation of landscape structure. 4D Tájépítészeti és kertművészeti folyóirat. J Landscape Archit Gard Art 36:32–39
Csorba P, Szabó Sz (2012) The application of landscape indices in landscape ecology. In: Tiefenbacher J (ed) Perspectives on nature conservation—patterns, pressures and prospects. pp 121–140. http://www.intechopen.com/books/perspectives-on-natureconservation-patterns-pressures-and-prospects/the-application-of-landscape-indices-in-landscape-ecology
De Groot RS, Alkemade R, Braat L, Hein L, Willemen L (2010) Challenges in integrating the concept of ecosystem services and values in landscape planning, management and decision making. Ecol Complex 6:453–462
Demek J (1974) Systémová teórie a štúdium krajiny. Studia Geographica, 40, Brno, Geografický ústav ČSAV, p 198
Demek J (1978) The landscape as a geosystem. Geoforum 9(1):29–34

Diviaková A (2011) Biotické komplexy pre environmentálny manažment. VKÚ, a. s., Harmanec, p 120
Drdos J, Urbánek J, Mazúr E (1980) Landscape syntheses and their role in solving the problems of environment, vol 32. Geografický Casopis, Bratislava pp 119–129
Drdoš J, Miklós L, Kozová M, Urbánek J (1995) Základy krajinného plánovania. Technická univerzita Zvolen, p 172
Ellenberg H (ed) (1973): Ökosystemforschung. Springer, Berlin, Heidelberg, New York
Farina A (1998) Principles and methods in landscape ecology. Chapmann and Hall, London
Feranec J, Oťaheľ J (2001) Krajinná pokrývka Slovenska/Land cover of Slovakia. Bratislava (Veda)
Finke L (1994) Landschaftsökologie. Westermann Verlag, Braunschweig
Forman RTT (1995) Land mosaics. The ecology of landscapes and regions. Cambridge University Press, Cambridge, p 632
Forman RTT, Godron M (1981) Patches and structural components for landscape ecology. Bioscience 31:733–740
Forman RTT Godron M (1986) Landscape ecology. Wiley, New York
Golley FB, Bellot J (eds) (1991) Interactions of landscape ecology, planning and design. Landscape Urban Plann 21(1–2):3–11
Grodzinski MD (2005) Piznannya landschaftu: misce i prostir (Understanding Landscape. Place and Space. In Ukrain). Kiev. U 2-x t. Vidavnicsho-poligrafichesky centr Kiivsky universitet. T.1, 431 pp., T.2, 503 pp
Grunewald K, Bastian O (eds) (2015) Ecosystem services—concept, methods and case studies. Springer-Verlag Berlin, Heidelberg, p 312
Gustafson EJ (1998) Quantifying landscape spatial pattern: what is the state of the art? Ecosystems 1:143–156
Guth J, Kučera T (1997) Monitorování změn krajinného pokryvu s využitím DPZ a GIS. Příroda 10:107–124
Haase G (1973) Zu Inhalt und Terminologie der topischen und chorischen Landschaftserkundung. Beitrag zu IIIrd Internacional Symposium on Problems of Ecological Landscape Research, Smolenice, Bratislava: ÚBK SAV, p 19
Haase G (1978) Zur Ableitung und Kennzeichnung von Naturraumpotentialen. Petermanns Geogr Mitt 122(2):113–125
Haase G (1980) Izučenie topičeskich i choričeskich struktur, ich dinamiki i razvitija v landšaftnych sistemach. In: Structura, dinamika i razvitije landšaftov. Moskva: Institut geografii AN SSSR, pp 57–81
Haase G et al (1991) Naturraumerkundung und Landnutzung. Geochorologische Verfahren zur Analyse, Kartierung und Bewertung von Naturräumen. Beiträge zur Geographie. Berlin, 34, I und II
Haase G (1996) Geotopologie und Geochorologie — Die Leipzig-Dresdner Schule der Landschaftsökologie. In: Haase G, Eichler E (Hrsg): Wege und Fortschritte der Wissenschaft. — Sächsische Akademie der Wissenschaften zu Leipzig, Akademie Verlag, pp 201–229
Haber W (1980) Entwicklung und Probleme der Kulturlandschaft im Spiegel ihrer Ökosysteme. Forsiarchiv Jg. 51, H. 12, 245–250
Haber W (2002) Kulturlandschaft zwischen Bild und Wirklichkeit. In: Schweizerisches Akademie der Geistes und Sozialwissenschaften. Bern, Akadenmievorträge, Heft IX, p 19
Haber W (2004) Landscape ecology as a bridge from ecosystems to human ecology. Ecol Res 19:99–106
Haber W (2005) Pflege des Landes – Verantwortung für Landschaft und Heimat. In: Deutscher Rat für Landespflege (ed): Landschaft und Heimat. Schriftenreihe des Deutschen Rates für Landespflege, Meckenheim 77:100–107
Haber W (2007) Vorstellungen über Landschaft. In: Busch B (ed) Jetzt ist die Landschaft ein Katalog voller Wörter. In: Beiträge zur Sprache der Ökologie, 5:78–85

Hofierka J, Šúri M (1996) Modelling spatial and temporal changes of soil water erosion. Geografický časopis 48:255–269

Hrnčiarová T, Miklós L (1991) Morphometric indices in the interpretation of water and material motion dynamics illustrated on the example Dolná Malanta. Ecology (CSFR) 10(2):187–221

Hunziker M, Buchecker M, Hartig T (2007) Space and place—two aspects of the human-landscape relationship. In: Kienast F, Wildi O, Ghosh S (eds) A changing world. Challenges for landscape research, Landscape Series, vol 8. Springer, Dordrecht, pp 47–62

Hynek A (2010) Krajina: objekt nebo konstrukt? In: Herber V (ed) Fyzickogeografický sborník 8. Fyzická geografie a kulturní krajina. MU, Brno, pp 138–142

Hynek A (2011) Geografie – geograficita prostorovost. In: Svobodová H (ed) Prostorovost: místa, území, krajiny, regiony, globiony. Sborník príspěvku z konference. GaREP, s.r.o. Brno, pp 6–50

Isachenko AG (1980) Optimization of the Natural Environment. Moscow: Mysl. Topological Aspects of the Science of Geosystems. Novosibirsk: Nauka, 1974

Isachenko AG (1981) Predstavlenije o geosisteme v sovremennoj fizičeskoj geografiji. Izv. VGO Leningrad 113(4):297–306

Iverson AL, Marín LE, Ennis KK, Gonthier DJ, Connor-barrie BT, Remfert JL, Cardinale BJ, Perfecto I (2014) Do polycultures promote winâ wins or tradeâ offs in agricultural ecosystem services? A metaâ analysis. J Appl Ecol 51(6):1593–1602

Jones KB, Zurlini G, Kienast F, Petrosillo I, Edwards T, Wade TG, Li B-L, Zaccarell N (2013) Informing landscape planning and design for sustaining ecosystem services from existing spatial patterns and knowledge. Landscape Ecol 28:1175–1192

Jongman RHG, Pungetti G (2004) Ecological networks and greenways: concept, design, implementation. United Kingdom at the University Press, Cambridge, p 153

IX. Kárpát-medencei környezettudományi konferencia (2013) Symposium abstracts, Miskolci Egyetem, Sapientia Erdélyi Magyar Tudományegyetem. Miskolc

Kerényi A (2003) Környezettan. Mezőgazda Kiadó, Budapest, p 470

Kertész Á (2002) Tájökológia. Holnap Kiadó, Budapest, p 166

Kertész Á (2010) Tájökológiai kutatások 2010. IV. Magyar Tájökológiai Konferencia, MTA FKI, Budapest, p 294

Khoroshev AV, Eremeeva AP, Merekalova K A (2013) Evaluation of intercomponents linkages in the steppe and taiga landscapes with account for modifiable areal unit. Proc of Russ Geogr Soc 145(3):32–42 (in Russian)

Kienast F, Wildi O, Ghosh S (eds) (2007) A changing world. Challenges for landscape research. Landscape series, vol 8. Springer, Dordrecht, pp 753–754

Kistowski (1998) Relations between the structure of landscape and its potential at the territory of last glaciation in Poland. In: The problems of landscape ecology, vol III, Warsaw, pp 179–188

Kolejka M et al (2011) Krajina Česka a Slovenska v současném výzkumu. Masarykova univerzita, Pedagogicka fakulta, Brno

Konkoly-Gyuró E, Tirási Á, Wrbka T, Prinz M, Renetzeder C (2010) Der Charakter grenzüberschreitender Landschaften. Das Fertő/Neusiedlersee-Hanság-Becken und die Region Sopron. Verlag Universität Westungarn. Lőverprint, Sopron, p 43

Kozová M, Hrnčiarová T, Drdoš et al (2007) Landscape ecology in Slovakia. Development, current state, and perspectives. Contribution of the slovak landscape ecologists to the IALE world congress 2007 and to the 25th anniversary of IALE. Bratislava: Ministry of the environment of the Slovak Republic, Slovak association for landscape ecology—IALE-SK, 2007, CD ROM, p 541

Krcho J (1968) Prírodná časť geosféry ako kybernetický systém a jeho vyjadrenia v mape. Bratislava, Geografický časopis, 20(2):115–130

Krcho J (1974) Štruktúra a priestorová diferenciácia fyzicko -geografickej sféry ako kybernetického systému. Bratislava, Geografický časopis 26(2):132–162

Krcho J (1978) The spatial organisation of the physical-geographical sphere as a cybernetic system expressed by means of measure as entropy. In: Acta Facultatis Rerum Naturalium Universitas Comenianae, Geographica, vol 16. pp 57–147

Krcho J (1991) Georelief as a subsystem of landscape and the influence of morphometric parameters of georelief on spatial diferentation of landscape-ecological processes. Ecology (CSFR), Bratislava 10(2):115–158

Landscape Ecology (2015) Symposium abstracts. In:17th International symposium. ILE SAS, Bratislava

Landscape Ecology: From Theory to Practice (2012) In: Proceedings of the 16th international symposium on problems of landscape ecological research. ILE SAS, Bratislava

Leser H (1991) Landschaftsökologie: Ansatz, Modelle, Methodik, Anwendung. Eugen Ulmer Verlag, Stuttgart, p 647

Leser H (1997) Landschaftsökologie. Ulmer, Stuttgart

Longatti P, Dalang T (2007) The meaning of "Landscape"—an exegesis of Swiss Government texts. In: Kienast F, Wildi O, Ghosh S (eds) A changing world. Challenges for landscape research. Landscape series, vol 8. Springer, Dordrecht, pp 47–62

Łowicki D, Mizgajski A (2013) Typology of physical-geographical regions in Poland in line with land-cover structure and its changes in the years 1990–2006. W: Geographia Polonica, Warszawa, 86(3):255–266

Mander Ü, Müller F, Wrbka T (2005) Functional and structural landscape indicators: upscaling and downscaling problems. Ecol Ind 5(4):267–272

McGarigal K (2002) Landscape pattern metrics. In: El-Shaarawi AH, Piegorsch WW (eds) Encyclopedia of environmetrics, Chichester, vol 2. Wiley, England, pp 1135–1142

Mezősi G, Fejes Cs (2004) Landscape metrics (in Hungarian), In: Táj és Környezet, MTA FKI, Budapest, Hungary, pp 229–242

Mičian Ľ (1982) A system approach to landscape and the sciences realizing it with a special to the system of geographical scienecs. In: VIth International symposium on problems of landscape ecological research, Piešťany, Institute of Experimental Biology and Ecology SAS, Bratislava, Oct 1982

Miklós L (1988) Space and position—scene of the origin of spatial ecological landscape problems. Ekológia (ČSSR) 7:381–395

Miklós L, Hrnčiarová T et al (eds) (2002) Atlas krajiny Slovenskej republiky. 1. vydanie, MŽP SR Bratislava, SAŽP Banská Bystrica, p 344

Miklós L, Izakovičová Z (1997) Krajina ako geosystém. VEDA, SAV, Bratislava, p 152

Miklós L, Špinerová A (2011) Krajinno-ekologické plánovanie LANDEP. VKÚ, Harmanec, p 159

Miklós L, Diviaková A, Izakovičová Z (2011a) Ekologické siete a územný systém ekologickej stability. Vydavateľstvo TU vo Zvolene, p 141

Miklós L, Ivanič B, Kočický D (2011b) Krajinnoekologická základňa integrovaného manažmentu povodia Ipľa. Digitálna databáza a tematické mapové vrstvy. Projekt HUSK 0801/2.1.2/0162 Vytvorenie jednotného monitoringu na báze priestorového informačného systému v povodí Ipľa. Banská Štiavnica, Esprit, spol. s r. o. (elektronický zdroj)

Miklós L, Izakovičová Z, Kanka R, Ivanič B, Kočický D, Špinerová A, David S, Piscová V, Štefunková D, Oszlányi J, Ábrahámová A (2011c) Geografický informačný systém povodia Ipľa: Katalóg GIS a výber máp. Bratislava: Ústav krajinnej ekológie SAV: Katedra UNESCO, Fakulta ekológie a environmentalistiky, Technická univerzita Zvolen, Esprit Banská Štiavnica, p 143

Miklós L, Németh R, Verrasztó Z (2014) Application of GIS in studying the drainage basin of the Ipoly River. Scientific Annals of the Danube Delta Institute, Tulcea, Romania, 20:109–128

Mizgajski A, Markuszewska I (eds) (2010) Implementation of landscape ecological knowledge in practice. The problems of landscape ecology. Volume XXVIII. Polish Association for Landscape Ecology, Wydawnictwo Naukowe Adam Miczkiewicz University, Poznań

Mizgajski A, Stępniewska M (2012) Ecosystem services assessment for Poland—challenges and possible solutions. Ekonomia i Środowisko, Białystok 2(42):54–73

Mosimann T (1984) Methodische Grundprinzipien für die Untersuchung von Geoökosystemen in der topologischen Dimension. In: Geomethodica, Veröff. 9. Basler Geometh. Coll., Bd. 9:31–65

Mosimann T (1990) Ökotope als elementare Prozesseinheiten der Landschaft (Konzept zur prozessorientierten Klassifikation von Geosystem). Geosynthesis, 1, Geographisches Institut der Universität Hanover, p 56
Nassauer JI (ed) (1997) Placing nature: culture and landscape ecology. Island Press, Washington, D.C
Naveh Z (1990) Landscape ecology as a bridge between bioecology and human ecology. In: Svobodova H (ed) Cultural aspects of landscape. Pudoc Wageningen, p 173
Naveh Z, Lieberman AS (1984) Landscape ecology—theory and application. Springer, New York, p 356
Naveh Z, Lieberman AS (1993) Landscape ecology—theory and application, 2nd edn. Springer-Verlag, New York
Naveh Z, Lieberman AS (1994) Landscape ecology: theory and application. Springer, New York
Neef E (1963) Topologische und chorologische Arbeitsweisen in der Landschaftsforschung. PGM 107(4):249–259
Neef E (1967) Die theoretischen Grundlagen der Landschaftslehre. H. Haack, Gotha, Leipzig
Neef E, Richter H, Barsch H, Haase G (1973) Beitrage zur Klarung der Terminologie in der Landschaftsforschung Geographisches Institut der Akademie der Wissenschaften der DDR, Leipzig, p 28
Odum EP (1975) Ecology: the link between the natural and the social sciences, 2nd edn. Holt Rinehart and Winston, London, New York, Sidney, Toronto
Oťaheľ J (2004) Landscape and landscape research in Slovakia. Belgeo - Revue belge de géographie 2–3. Landscape research in Europe
Oťaheľ J, Feranec J, Cebecauer T, Pravda J, Husár K (2004) Krajinná štruktúra okresu Skalica: hodnotenie zmien, diverzity a stability. Geographia Slovaca, 19, p 123
Pedroli B (ed) (2000) Landscape our home (Lebensraum Landschaft). Indigo, Zeist Stuttgart
Preobrazhensky VS (1983) System orientation of landscape research in geography and its present-day realization. In: Drdoš J (ed) Landscape synthesis. Geoecological foundations of the complex landscape management. VEDA, Bratislava, pp 31–36
Preobrazhensky VS, Minc AA (1973) Sootnoshenye ponyaty geosystema a ekosystema. In: Práce a materiály z biológie krajiny 20. Proceedings of 3rd international symposium on the landscape ecological research. ÚBK SAV, Bratislava
Proceedings of 3rd International Symposium on the Landscape Ecological Research (1973) Práce a materiály z biológie krajiny 20, ÚBK SAV, Bratislava
Proceedings of 4th Congress of the Czechoslovak Geographers (1978) Geographical institute of the SAS, Bratislava
Proceedings of 4th International Symposium on the Landscape Ecological Research (1976) Institute of experimental biology and ecology SAS, Bratislava
Proceedings of 5th International Symposium on the Landscape Ecological Research (1979) Institute of experimental biology and ecology SAS, Bratislava
Proceedings of 6th International Symposium on Problems of Landscape Ecological Research (1982) Institute of experimental biology and ecology SAS, Bratislava
Puzachenko JG (2006) Landscape planning: general fundamentals, methodology, technology. Moscow, pp 4–80 (in Russian)
Renetzeder C, Wrbka T, Mücher S, Eupen MV, Kiers M (2010) Landscape structure analysis in selected european regions and its use for sustainability impact assessment. In: Andel J, Bicik I, Dostal P, Lipsky Z, Shahneshin SG (eds) Landscape modelling. Geographical space, transformation and future scenarios, Urban and landscape perspectives, vol 8, p 236
Richling A, Solon J (1993) Ekologia krajobrazu. Wydaznictvo Naukowe PWN, Warszawa, p 225
Richling A, Solon J (1996) Landscape ecology. Wydawnictwo Naukowe PWN, Warszawa, p 464 (in Polish)
Risser PG, Karr JR, Forman RTT (1984) Landscape ecology: directions and approaches. Illinois Natural History Survey. Special Publ. 2, Champaign

Ružička M (2000) Krajinnoekologické plánovanie - LANDEP. (Systémový prístup v krajinnej ekológii). Združenie BIOSFÉRA. Bratislava
Ružička M, Miklós L (1982) Landscape ecological planning (LANDEP) in the process of territorial planning. Ekologia (CSSR) 1(3):297–312
Ružička M et al (1978) Krajinné zložky, prvky a štruktúra v biologickom plánovaní krajiny. Quasestiones geobiologicae 23:7–63
Ružičková H, Ružička M (1973) Druhotná štruktúra krajiny ako kritérium biologickej rovno¬váhy. Quaestiones geobiologicae 12:23–62
Schilleci F, Todaro V, Lotta F (2017) Connected lands: new perspectives on ecological networks planning. Springer, p 85
Schmithüsen J (1968) Was ist eine Landschaft?. Universität d. Saarlandes, Saarbrücken
Schmithüsen J (1974) Landschaft und Vegetation. Geograph. Inst. d. Univ. d. Saarlandes, Saarbrücken
Schmithüsen J (1976) Allgemeine Geosynergetik. Grundlagen der Landschaftskunde. de Gruyter, New York, Berlin
Schreiber KF (1980) Brachflachen in der Kulturlandschaft. Daten und Dokumente zum Umweltschutz, Sonderreihe Umwelttaguug 'Okologische Probleme in Agrarlandschaften', Univ. Hohenheim 30:61–93
Schreiber K F (1985) Was leistet die Landschaftsäkologie für eine àkologische Planung? Schriften reihe für Orts-, Regional-und Landesplanung 34. Zürich, Switzerland, pp 7–28
Snacken F, Antrop M (1983) Structure and dynamics of landscape systems. In: Drdoš J (ed) Landscape synthesis: geoecological foundations of the complex landscape management. VEDA, Bratislava, pp 10–30
Snytko VA (1983) Substance dynamics in geosystems GeoJournal 7(2):135–138
Snytko V, Semenov YM (2008) The study of geosystem structure, development and functioning in Siberia. The V. B. Sochava Institute of Geography, Siberian Branch of RAS Irkutsk
Sochava VB (1977) Landscapes of Southeastern Siberia (1:1 500000 map). Russia, Moscow
Sochava VB (1978) Vvedenie v učenije o geosistemach (An Introduction to the Science of Geosystems in Russian). Nauka, Novosibirsk, p 319
Špinerová A (2015) Štruktúra krajiny ako regulátor dynamiky pohybu vody a materiálu. Technická univerzita Zvolen, p 132
Špulerová J, Dobrovodská M et al. (2011) Stratégia ochrany a manažmentu historických štruktúr poľnohospodárskej krajiny v modelovom území Liptovská Teplička. Bratislava: Ústav krajinnej ekológie SAV, p 152
Stanová V, Valachovič M et al (eds) (2002) Katalóg biotopov slovenska. Bratislava: DAPHNE – Inštitút aplikovanej ekológie, p 225
Šteffek J, Gallay I, Gallayová Z, Kunca V, Lepeška T, Olah B, Škvareninová J, Ujházy K, Ujháziová M, Wiezik M, Križanová Z (2008) Krajinnoekologický výskum: teória, metódy a prax. Zvolen: FEE TU, p 222
Štefunková D et al (2011) Vybrané výsledky výskumu starej vinohradníckej krajiny Svätého Jura. In Ekologické štúdie: Vedecký časopis venovaný aktuálnym problémom ekológie, krajinnej ekológie a príbuzných vedných disciplín 2(1–2):126–131, ISSN 1338-2853
Štefunková D, Hanušin J (2015) Analysis of the spatial and temporal distribution of selected landscape diversity indexes in detailed scale (Example of Viticultural Landscape Svätý Jur). In: Halada Ľ, Bača A, Boltižiar M (eds) Landscape and landscape ecology. Proceedings of the 17th international symposium on landscape ecology. Institute of landscape ecology, Slovak Academy of Sciences, Bratislava, pp 185–191
Szabó M (2007) Tájszerkezeti változások a Felső-Szigetközben az elmúlt 20 évben. Földrajzi Közlemények. CXXXI. (LV). 1(2):55–74
Szabó M (2008) A bioszféra sokfélesége—az élővilág változatossága és veszélyeztetettsége. Földrajzi Közlemények. 132(4):441–449

Szabó Sz, Csorba P, Varga K (2008) Landscape indices and landuse—tools for landscape managment. In: Plit J, Andreychouk V (eds) Methods of landscape research, Dissertations Commission of Cultural Landscape, Sosnowiec, Poland, 8:7–20, ISSN 1896-1460

Szádeczky-Kardoss E (1989) A jelenségek univerzális kapcsolódása (Akadémiai Kiadó Bp., pp 1–292)

Tansley AG (1935) The use and abuse of vegetational terms and concepts. Ecology 16(3):284–307

Tarboton DG (1997) A new method for the determination of flow directions and upslope areas in grid digital elevation models. Water Resour Res 33(2):309–319

Teleki P (1917) A földrajzi gondolat története -MTA székfoglaló, (Kossuth Könyvkiadó. Bp. 1996:1–194

Tjallingii SP, De Veer AA (eds) (1982) Perspectives in landscape ecology. In: Proceedings of the 1st international congress in landscape ecology, Veldhoven, The Netherlands, Center for Agricultural Publishing and Documentation, Wageningen, 6–11 Apr 1981

Tremboš P (1994) Geoekologická syntéza Diviackej pahorkatiny - metodický postup vymedzovania základných krajinných jednotiek. In: Minár, J. (ed.): Prírodná časť krajiny, jej výskum a návrhy na využitie, /Zborník/, KFGaG PRIF UK, Bratislava, pp 118–120

Tress B, Tress G (2001) Capitalising on multiplicity: a transdisciplinary systems approach to landscape research. Landscape Urban Plann 57(3–4):143–157

Troll C (1939) Luftbildplan und ökologische Bodenforschung (Aerial photography and ecological studies of the earth). Zeitschrift der Gesellschaft für Erdkunde, Berlin, pp 241–298

Turner M (1990) Spatial and temporal analysis of the landscape patterns. Landscape Ecol 4(1):21–30

Turner M G, Gardner RH (eds) (1991) Quantitative methods in landscape ecology: the analysis and interpretation of landscape heterogeneity. Springer

Vejre H, Brandt J (2004) Contemporary Danish landscape research. Belgeo—Revue belge de géographie 2–3. Landscape research in Europe

Verrasztó Z (1979) Land formation and the geological aspects of environmental protection. In: Symposium Changes of the geological environment under the influence of man's activity. IAEG National group, Krakow-SandomierzBelchatow-Plock-Warszawa, pp 135–141

Verrasztó Z (2017) Holisztikus környezeti modellezés kartográfiai, GIS és távérzékelési módszerek alkalmazásával. Remote Sens 7(1):490–516

Veteikis D, Kavaliauskas P, Skorupskas R (2015) Assessing the optimality of landscape structure in a landscape plan (a Lithuanian example) In: Halada Ľ, Bača A, Boltižiar M (eds) Landscape and landscape ecology. Proceedings of the 17th international symposium on landscape ecology. Institute of Landscape Ecology, Slovak Academy of Sciences, Bratislava, pp 348–358

Vidal De La Blache P (1922) Principes de géographie humaine. Armand Colin, Paris

Vološčuk I (2003) Ochrana prírody krajiny. TU vo Zvolene, Zvolen

von Bertalanffy L (1950) An outline of general system theory. Br J Philos Sci 1(2):134–165

von Bertalanffy L (1968) General system theory. Foundations, development and applications. George Brazileer, Penguin Books, New York

Wascher D (ed) (2005) European landscape character areas. Alterra Report, No. 1254, p 160

Wascher D, Jongman R (2000) European landscapes, classification, assessment and conservation. European Environmental Agency, Coppenhagen

Wrbka T (2009) Das Waldviertel - Wildnisreste in historisch gewachsener Kulturlandschaft am Grünen Band Europas. In: Wrbka T, Zmelik K, Grünweis FM (eds) Das Grüne Band Europas - Grenze.Wildnis.Zukunft. Weitra. Bibliothek der Provinz, p 342

Wrbka T, Erb KH, Schulz NB, Peterseil J, Hahn CO, Haberl H (2004) Linking pattern and process in cultural landscapes. an empirical study based on spatially explicit indicators. Land Use Policy 21:289–306

Wrbka T, Stocker-Kiss A, Schmitzberger I, Peterseil J (2005) Network landscapes of Austria—identification and ecological assessment. In: Konkoly Gyuró É (ed) Greenways. Conference presentations on ecological corridors, green corridors, Sopron, Hungary, pp 47–62

Wu J (2013) Key concepts and research topics in landscape ecology revisited: 30 years after the Allerton Park workshop. Landscape Ecol 28:1–11

Wu J, Hobbs R (2002) Key issues and research priorities in landscape ecology: an idiosyncratic synthesis. Landscape Ecol 17:355–365

Žigrai F (1983) Krajina a jej využívanie. 1. vyd. Brno: UJEP, 1983, p 131

Žigrai F (1995) Integračný význam štúdia využitia zeme v geografii a krajinnej ekológii na príklade modelového územia Lúčky v Liptove. Geografické štúdie, p 4

Žigrai F (2015) Meta-landscape ecology as a new ecological science (selected metascientific aspects). In: Halada Ľ, Bača A, Boltižiar M (eds) Landscape and landscape ecology. Proceedings of the 17th international symposium on landscape ecology. Institute of landscape ecology, Slovak Academy of Sciences, Bratislava, pp 25–35

Zonneveld JIS (1981) The ecological backdrop of regional geography. In: Tjallingii SP, de Veer AF (eds) Perspective in landscape ecology. Proceedings of the international congress of the Netherlands society of landscape ecology, Veldhoven, PUDOC, Wageningen, pp 67–90

Zonneveld IS (1995) Land ecology: an introduction to landscape ecology as a base for land evaluation. Land management and conservation. SPB Academic Publishing, Amsterdam

Chapter 3
Construction and Mapping of Geocomplexes

Abstract The chapter introduces the application of the geosystem theory for the creation of practically usable, complex spatial landscape-ecological units—geocomplexes of different character. The content and characteristics of these units represent the spatial/contentual frame for geosystem based information system on the landscape. The basic steps for creation and spatial design of these spatial units are the landscape-ecological analyses and landscape-ecological syntheses. The result of this procedure is the definition of the content, assignation and spatial display of geocomplexes of different complexity in the following range: abiotic complex → physical-geographical complex → complex of land cover/land use and real vegetation → landscape-ecological complex → socio-economic complex → integrated geocomplex. The analyses and syntheses are of character of basic research on the landscape, as well as the initial step to the applied procedures and planning practice. The use of a proper information system for the systematic characteristics of the geocomplexes is an inevitable necessity. The chapter introduces the methodical and practical problems of the creation of such database within a GIS frame. The synthetic, well defined and precisely mapped geocomplexes are considered the **basic operational units** of the decision-making process for integrated landscape management. The final part of the chapter describes the real situation of the creation and problems of the database of geocomplexes in Slovakia. The text is completed with numerous tables which illustrate the most frequently used state variables and typological characteristics of different complexes.

Keywords Geocomplex · State variables · Typological characteristics · GIS Slovakia

Considering the objective of specific landscape-ecological works it is reasonable to work with the concept of **geosystems** in terms of their understanding as **geocomplexes** (see Chap. 2.3). As defined, the geocomplexes are the same material objects as geosystems, have the same elements but in their models the relation within the geosystems are not explicitly described.

L. Miklós et al., *Landscape as a Geosystem*,
https://doi.org/10.1007/978-3-319-94024-3_3

The values of single indicators of the properties of geocomplex elements have their own spatial distribution controlled by natural rules. They occur in certain combinations—in complexes—and on limited sites have **homogeneous values of all indicators**, distinct from adjacent homogeneous sites with different combinations of values—other content of the complex. It is a crucial factor for decision-making on optimal use of the landscape. Therefore, a basic effort of landscape-ecological studies is the

- recognition and definition of the **analytical content** of complexes (combination of values of indicators);
- **demarcation** of homogeneous areas with a defined combination—mapping of complexes.

The homogenous sites of complexes—the areas of geocomplexes—should divide the territory without remainder.

3.1 Framework Approach for Assignation and Delimitation of Geocomplexes

The basic aspects of geocomplexes include:

- they are defined by set of the indicators of properties and their values;
- within their boundaries all defined indicator has unchanging (homogeneous) values;
- sites of geocomplexes serve as a **spatial database** for further applications and approaches;
- spatial units of geocomplexes may have a character in the territory as repeatedly occurring **types** or not repeatedly individual **regions**;
- the set of values of analytical indicators in the given type of geosystem represents defined **equal conditions** for decision-making process on the optimal utilization of landscape, the homogenous area means an **equal suitability for use** by humans on the entire area of a given type and on all occurrences of this type of geosystem.

Thereby synthetic geocomplexes become **basic operational units** for the further **decision-making process**.

For creation of geocomplexes we can use the routine procedure of landscape-environmental analysis and landscape-ecological synthesis, developed through the methodology of landscape-ecological planning, LANDEP (Ružička and Miklós 1982, 1990; Miklós and Špinerová 2011). The next chapter will describe the process of construction of different types of geocomplexes through landscape-ecological analyses and syntheses, as well as their characteristics.

3.1.1 Landscape-Ecological Analyses

The basic task of landscape-ecological analysis in the LANDEP methodology is the acquisition and homogenisation of the information on the primary, secondary/current and tertiary/socio-economic structure of the landscape. Landscape-ecological analyses should play the role of **documentation** of the whole further process of syntheses. They are also thereby an input into the decision-making process for the optimal utilization of the landscape.

The content is a purpose-oriented review, homogenisation and modification of existing analytical documents from a wide variety of specialized research. Information is collected by field survey, excerption from literature sources, from various maps and statistical documentation, from aerial photography and consultation with experts for the given matter. Nevertheless, many special analyses are not available, they must be prepared specifically for the given task.

According to the above-mentioned, the objective of landscape-ecological analysis is the correct definition and determination of the values of indicators of properties of the landscape elements. As described in the Chap. 2.4, indicators of the properties of landscape elements can be defined as:

- **state variables**—real measurable values of individual indicators of properties of elements that determine the current state of the element (e.g. the depth of the soil, rainfall, flow, tree height, amount of biomass);
- **typological characteristics of spatial units**—spatial subsystems; These often have a narrative character, they express overall characteristics which reflect several properties of the given element; they are usually expressed in maps as spatial units (e.g. soil type, climate type, physiognomic-ecological formation of plants);
- **combination of typological characteristics and state variables**.

It should be noted again that it is impossible to lay down precise rules of which elements in specific case studies should be characterized by state variables and which by typological characteristics. With a large degree of generalization we can say that state values should be used in particular in landscape-environmental papers on a large scale and in specialized studies, while typological characteristics are popular in informative and descriptive studies, as well as less detailed works on a smaller scale.

Analyses have the following parts:

(a) **Defining the Territory of Interest**

This step is absolutely the first—practically the zeroth input into the process. It is appropriate to define a coherent territory, with natural borders that "wrap" administrative or economic boundaries. The decision will be more successful the larger and more diverse the area is, as it will provide a wider space for the optimal deployment of activities. This means that the ecological optimization of land use should continue from top to bottom.

For the area of interest it is suitable to prepare a simplified cartographic projection basis with the main topographical features. This cartographic basis should be used for the projection of all the working and resulting material. The projection of all processed information into a single unified base is a prerequisite for correct synthesis, as well as the successful completion of any additional steps of different works.

We divide the actual landscape-ecological analyses into

(b) **Analysis of the Primary Landscape Structure**

These analyses relate particularly to analyses of **abiotic elements of geosystems**. For practical purposes considering the physical character and relations of the elements of geocomplexes it is suitable to subdivide the primary landscape structure into partial complexes such as:

- *Substrate–soil–water complex*

The most frequently used indicators in this complex are: depth, skeletality, and grain size of the substrate and the soil, and depth of subsurface water (Table 3.1a);

- *complex relief—slope dynamics*: particularly slope angle, horizontal and normal curvature of the relief, direction of gradient curves, topographic position, slope length, zone dynamics of slope;
- *complex relief–(micro)climate*: particularly the orientation of relief to compass points, insolation, shading;
- *complex relief–dissection*: particularly morphological and morphometric types of relief, horizontal and vertical dissection of relief (examples see in Sect. 4.1).

The most frequently used indicators in this complex are shown on Table 3.1a.

As described in previous chapter a specific interpretation of the whole abiotic complex allows us to define the

- abiocomplex—potential natural vegetation (Table 3.1b).

A deeper insight into the values of indicators is provided by the on Tables 4.5–4.20 (Sect. 4.1).

(c) **Analysis of Secondary Landscape Structure**

In practice, we analyse and map the secondary landscape structure in its today form, as the current land cover, so for this structure is more illustrative the term **current landscape structure (CLS)**. CLS is formed from elements which mankind has partially or fully influenced—e.g. the forests, reshaped—e.g. the agricultural land, and newly created—e.g. the buildings and other technical objects, which currently cover the land's surface (Miklós and Izakovičová 1997). These elements can be characterized with different levels of detail according to their biological and ecological content as physiognomic-ecological formations, as communities of real vegetation, as well as habitats or biotopes. They have the character of topical units, and in this understanding we refer to them as CLS mapping units.

The basic division of CLS mapping units still retains the original division of the secondary landscape structure, as formulated by Ružičková and Ružička (1973) and

Table 3.1a Frequently used state variables and typological characteristics of the primary landscape structure: complex substrate–groundwater–soil and complex relief–dynamics of the surface–dissection–position

State variables	*Typological characteristics
Geological substrate: rocks, weathered rocks, sediments	
Texture and structure	Pre-quaternary geological formations (geological maps)
Resistance against weathering (geomorphological value)	Quaternary formations
Chemical structure (mineral strength)	Types of rocks, weathered rocks and covers: litological character
Engineering geological substrate value	
Permeability and aquifer	
Subsurface water	
Resource efficiency	Types of underground water according to permeability
Chemistry and mineralization	Types of underground water according to mineral composition
Depth of the underground water level according to above sea level	Types of underground water according to pressure ratios
Depth of the level under the terrain	
Regime of underground water	
Soil	
Depth and skeletality	Soil types and subtypes
Texture (soil group)	
Structure, consistency, waterlogging by surface water	
Water, salt and thermic regimes	
Humus and carbonates content	
Chemistry and soil sorption	
(Geo)relief	
(a) Topic (vertical) parameters of relief	Morphotopes
Coordinates	
Altitude and relative terrain heights	
Gradient of relief	
Horizontal and normal curvature	
Forms of relief	
Orientation of the relief to the cardinal points (aspect)	
(b) Choric (horizontal) parameters of relief	

(continued)

Table 3.1a (continued)

State variables	*Typological characteristics
Gradient curves–direction of gravity	
The same direction of gradient–dividing lines	Morphochores (elementary gravitation areas, microcatchments)
Length of the gradient curve	
Shape and size of elementary gravitation areas	
Integration mode of elementary gravitation areas	
(c) Topical-choric parameters of relief	Morphotopochores
Topographic location of morphotopes	Morphogenetic types of relief
Topographic location of relief types	Orographic units
Orographic position of relief types	Natural-positional regions
Vertical and horizontal segmentation of mezorelief	Morphographic-morphometric types of the relief
Total relief segmentation	Forms of relief: other morphometric spatial units (e.g. Anthropogenic forms of relief)
Relative elevation of morphotopes	
Relative elevation of microcatchments and depression	
Topographic location and microcatchments connection	Types of microcatchments by location
Neighbourhood form of microcatchments, distance of microcatchment from the end of hydrological system	

*Typological characteristics do not connect necessarily to the status variables in given row!

Ružička et al. (1978). The differences are only in the degree of detail of work, which is based on the objectives of the specific task and the overall character of the current landscape structure of the model territory (Table 3.2).

Individual economic sectors and other spheres of society's development make demands for their development on a certain area of the landscape, both existing and newly required areas. Interest is implemented that a given sector defines a certain area, changes its use, its structure, builds a building, defines the functional zone, and these areas are engaged in a specific activity. In other words, areas are required in order for the activities the sector needs to be carried out. These activities—which are naturally dynamic, often permanent—must be depicted on maps (plans and projects), which, of course, cannot capture the dynamics. Therefore, activities on maps—as a result of the decision-making process—are displayed statically not as activity, but as a **site of activity**, that is labelled as an element of the current landscape structure in the form of **functional elements of land use**, real elements of vegetation, habitats, etc.

Table 3.1b Frequently used state variables and typological characteristics of the primary landscape structure: complex relief—surface waters, complex relief—climate, abiocomplex—potential vegetation

State variables	*Typological characteristics
Complex relief—surface water	
Run-off coefficient and specific surface run-off	
Integration of run-off (quantity)	
River basin area	
Spatial structure and shape of hydrological systems	
Hierarchical division of hydrological systems	Individual river basins
River flow and the amount of run-off	
Drain regime and other river regimes (thermal, ice drift, sediments regime)	River types (by regime)
Complex relief-climate	
(a) Meteorological indicators	
Basic: temperature, rainfall, humidity, pressure (averages, extremes, duration, regime)	Climate regions and sub-regions
Purpose indicators: temperature inversion	Climate geographic types
Purpose indicators: torrential rain, specific expense	Agro-climatic areas
Wind conditions (speed, direction, duration)	Phenological-geographic types
(b) Morphoclimatic indicators	
Insolation of the georelief (duration, quantities of incident sunlight)	Morpho-clima-topes
Shading of the relief and running the relief shadow	Morpho-clima-chores
Coefficient of deceleration or acceleration of wind	Morpho-clima-topochores
Shading of relief against different landscape elements	
Complex abiotical subcomplex—potential vegetation	
Theoretically predicted species composition of vegetation	Geobotanical units
Ecological and physiognomic characteristics of vegetation formation	Phytogeographic areas
Theoretically predicted species composition of fauna	Regiotyps of fauna and zootesseras
State variables of other elements of geosystem (climate, water, soil, geological substrate, relief)	Typological units of other elements of geosystem (climate, water, soil, geological substrate, relief)

*Typological characteristics do not connect necessarily to the status variables in given row!

Table 3.2 Frequently used state variables and typological characteristics of the secondary landscape structure

State variables	Spatial subsystems
Complex of real vegetation cover	
Species and its characteristics	
Physiognomicaly ecological formation and its charakteristics	Physiognomically ecologically-functional-spatial units of real vegetation: mapping units of pls)
Species composition: floristic characteristics	
Grouping of species into communities: geobotanical characteristics	
Population characteristics: frequency, dynamics/density	
Ecological/site characteristics: humidity, soil response, trophic and climatic conditions	
Biostructural characteristics: life formes, shapes, cover, dimensions, vegetation layer, age, phenological phases, diversity, distribution of diaspore	
State of the system/state function: balance, synantrophy, hemorobhy	
Production characteristics: feeding potential, honey production potential, therapeutic effect, plant toxicity	
Chorological characteristics: site size and its variation in time	
Ekosozological charakteristics: rarity, vulnerability, endemits, relicts, protected species	
Complex of biotopes	
Physiognomico-ecological characteristics of ***biotopes of faunal community***	Biotopes of faunal community
Species composition and clusters within zoocenoses	
Population characteristics: abundance, diversity	
Characteristics of bioindicators	
Ecosozological *characteristics*	
Complex land use	
Land use formes	Elements of land use (land use, land cover)
Structural characteristics: size, shape of parcels	
Functional properties: position, availability, machinability	
Complex of urban-technical structures	
Functional-spatial characteristics: mode of the use of areas and objects	Urban-technical elements
Structural properties: shape, size, material, spatial expression	

We can divide the most frequently mapped CLS elements/sites of social activities according to the nature, dynamics, according to their expected impact on the landscape into several groups:

- Ecostabilizing activities (greenery in the landscape, forests, bodies of water);
- Agricultural activities (arable land of various basic uses, permanent grassland);
- Permaculture (vineyards, orchards, gardens, allotments);
- Recreational activities (groups of cottages, recreational areas);
- Investment activities (various construction)

A deeper insight into the values of indicators is provided by the Tables 4.21–4.23 (Sect. 4.2).

(d) **Analysis of Tertiary Landscape Structure**

In practice, we are mapping the **socio-economic factors/phenomena** in the landscape (SEF). SEF are non-material, intangible in nature, they exist only as different standards, regulations and acts. Nevertheless, they can be mapped, more accurately projected onto the map, where they overlap each other and display the conflicts of interest in the landscape. We analyse both positive SEF—particularly SEF for protection of nature and resources, as well as SEF of a negative character—stress factors (Table 3.3).

A large part of the mapped SEF is firmly bound to the tangible elements or complexes of the primary, and, in particular to secondary landscape structure (Izakovičová et al. 2000). A deeper insight to this relation is given on Tables 3.4a, 3.4b, 3.4c, 3.4d, 3.4e, and on Tables 4.24–4.27 (Sect. 4.3).

3.1.2 Landscape-Ecological Syntheses

As described above, the values of single indicators of the properties of geocomplex elements have their own spatial distribution controlled by natural rules. For the practical use all analytical indicators of landscape properties and their values are parameterised, thus characterized by numbers–codes, including the properties that actually have verbal qualitative characteristics.

Of course, the choice of analysed properties varies for different areas and for different tasks. Important criteria for analysis are:

- indicators should convey diagnostic properties, they can be **parameterised and coded**;
- they can be **mapped** across the whole territory of interest;
- they **differentiate** the territory of interest, i.e. the affected area has varying values.

Less important are indicators that have the same value throughout the territory, thus do not differentiate territory.

Analyses are usually expressed on maps with coded values of indicators of landscape properties and serve as the basis of synthesis.

Table 3.3 Groups and typological characteristics of the tertiary landscape structure

SEF group	SEF content	SEF spatial manifestation
(a) SEF of nature conservation	Territorial nature protection	Protected area, protective zone
	Habitat of protected animals and plants	Protected area, protective zone
	Protection of species and individuals	Point (object), protective zone
(b) SEF of protection of natural resources	Protection and use of water resources	Protective zone/hygiene protection zone of an object (point), section
	Protection and use of natural healing resources	Protective zone/hygiene protection zone of an object (point), section
	Protection and use of forest resources	Site, USDF (unit of spatial division of forest), area
	Protection and use of soil resources	Parcel, site, administrative territory
	Protection and use of mineral resources	Site, point (object, borehole)
	Protection and use of cultural-historical resources	Structure, site, protective zone of site
(c) SEF of urban, residential and recreation areas	Urbanistic units and functional zones	Zone
	Recreational areas and zones	Zone, site, selected territory
(d) SEF of communication and transport	Protection and safety of transport structures	Protective/safety zones
	Protection and safety of energy/product pipelines	Protective/safety zone
(e) SEF of industrial and mining activities	Protection and safety of municipal facilities and sites (waste water treatment, waste management)	Protective/safety zone
	Protection and safety of storage facilities and sites	Protective/safety zone
	Protection and safety of production and mining facilities and sites	Protective/safety zone
(f) SEF of agricultural activities	Protection and safety of agricultural-technical facilities and sites	Protective/safety zone
	Protection of health in range of facilities and sites of animal production	Hygienic zones
(g) SEF of territorial units—administrative boundaries	State-administrative and regional units	Territorial boundary
	Sectorial territorial units	Territorial boundary
(h) SEF of deterioration of natural resources and the environment	Declared territory with polluted environment.	Borders, areas, sections, points and sites

Table 3.4a Relationship of elements of the secondary landscape structure: SEF related to the landscape as a whole and greenery

Secondary landscape structure: elements of land cover		Tertiary landscape structure: socio-economic factors/phenomena	
Landscape as a whole	Specific landscape structure	Protected landscape area	Territorial nature protection
		National park and buffer zone	
		Protected site and buffer zone	
		Natural reserve and buffer zone	
		Nature monument and buffer zone	
		Private protected area	
		Area of the NATURA 2000, Sites of Community Importance protected bird area	Other systems of nature protection
		Biotopes of Community Importance	
		Biosferic reservation MAB, UNESCO	
		The UNESCO World Cultural and Natural Heritage Site	
		Protected landscape area	
		National park and buffer zone	
		Protected site and buffer zone	
		natural reserve and buffer zone	
		Nature monument and buffer zone	
		Private protected area	
	Wetlands	Protected wetland ecosystems included in Ramsar convention	

(continued)

Table 3.4a (continued)

Secondary landscape structure: elements of land cover		Tertiary landscape structure: socio-economic factors/phenomena	
Mosaic of woody and grassland vegetation	Floodplains, groves	Biologically and aesthetically important area	
	Scrubs	Historical landscape structures	
	Lanes and line trees	Threatened valuable landscapes	
		Biocentres, biocorridors, interacting elements	TSES, protected biotopes
	Corresponding elements of SLS	Protected plant	Species protection
		Protected animal	
		Protected mineral	
		Protected fossil	
	Solitaires	Protected tree	

The values of single indicators of the properties of geocomplex elements occure in certain combinations—in complexes—and on limited sites have **homogeneous values of all indicators**, distinct from adjacent homogeneous sites with different combinations of values—other content of the complex. Accordingly, the step of 'synthesis' in the LANDEP methodology involves the formation, description and classification of homogeneous areas of geocomplexes of various ranges with a clearly defined set of analytical properties, and the segmentation of the study area into these areas.

The landscape understood as a geosystem there are two basic types of bonds, **vertical** and **horizontal**. Accordingly, the process the synthesis can be divided into typification and regionalization.

- in landscape-ecological **typification** according to vertical (topical) relations created spatial units with precisely defined homogeneous content in a given area have the same values of selected properties repeated at each occurrence. These are the types of abiocomplexes, physical-geographical complexes, landscape-ecological complexes (see below);
- landscape-ecological **regionalization** is a synthesis along horizontal (choric) relations aimed at creating hierarchically higher spatial and structural unit of a choric character—regions. They are spatial subsystems of different levels according to the characteristic spatial and content groupings of lower hierarchical units. The smallest are the elementary gravitation areas on the basis of the direction of gravitation, through microcatchment areas (Hrnčiarová and Miklós 1991; Miklós et al. 1984), river basins, up to natural-settlement gravitational **regions** (Miklós and Špinerová 2011).

Table 3.4b Relationship of elements of the secondary landscape structure to the tertiary landscape structure: SEF related to forests—waters—soils

Secondary landscape structure: elements of land cover		Tertiary landscape structure: socio-economic factors/phenomena	
Forests	Coniferous forests Broad leaved forests Mixed forests Monocultures	Commercial forest	Forests sources
		Forest nurseries	
		Protection forests	
		Forest for specific purposes	
Waters	Wells, groundwater resources	Protection zones of wells	Water sources
		Protected groundwater area	
	Water flows, catchment areas	Water catchment area	
		Protection zones of water reservoirs	
	Water flows, lakes, water reservoirs	Water suitable for bathing	
		Water suitable for irrigation	
		Waters suitable for fish and for reproduction of original species	
		Fishing grounds, fish habitat protected area	
	Inundated alluviums, areas in between dikes, polders	Declared inundation territory	
Other water elements	Thermal springs and other waters with specific properties	Protection zones of natural healing and mineral waters	Healing water sources
		Natural healing baths and their protection zones	
		Natural spa areas and their protection zones	
		Natural healing sources of gases and other emanations and their protection zones	

(continued)

Table 3.4b (continued)

Secondary landscape structure: elements of land cover		Tertiary landscape structure: socio-economic factors/phenomena	
	Peatlands, poor fen	Natural healing sources of peatlands, poor fens, muds and other soils and their protection zones	
Arable land and other agricultural elements	All types of agriculture	Parcels, registered culture in the real estate register culture,	Soil sources
	Arable land	Bonitated soil-ecological unit (BPEJ)	
		Drained and irrigated fields	
	Permanent grasslands		
	Orchards, vineyards, hop gardens	Registered areas	
	Areas of greenhouses	Investment of permanent character	
	Gardening areas		

Most often in practice, the methods of traditional physical-geographic typification are applied, using the method of lead indicator—most often the relief, geology or soils. In the process of landscape-ecological **regionalization**, the types of complexes are associated in accordance with predetermined relations to less homogeneous spatial units, each unit formed has an individual character. Selected properties upon which the regions are created, have such a combination that in a given territory, each individual region occurs only once.

(a) **Topical Synthesis—Formation of Topical Complexes (Abiocomplexes, Landscape-Ecological Complexes) and Their Types**

The most commonly used method of such synthesis is the **method of superposition** of analytic maps, resulting in a parametric typified area–creating maps of the types of abiotic complexes (ABC), biotic-anthropogenic complexes (BAC), socio-economic complexes (SEC) or landscape-ecological complexes (LEC). Currently, for this process is often used the technique of **geographical information systems** (GIS).

The method of the **leading indicator** lies on a preference of an indicator which has high spatial differentiation, is easily observable, can be achieved in a credible manner and there are significant correlations between the spatial differentiation of this feature and the spatial differentiation parameters of other considered elements. All the stated preconditions are best met by georelief. Georelief is one of the deci-

Table 3.4c Relationship of elements of the secondary landscape structure to the tertiary landscape structure: SEF related to recreation—municipalities—transport

Secondary landscape structure: elements of land cover		Tertiary landscape structure: socio-economic factors/phenomena	
Recreation elements	Camping sites	Declared zones of campgrounds	Recreation areas
	Cottages area	Cottage settlements	
	Recreational centres	Declared recreation areas, zones and centres of various level	
	Spa and sanatoriums	Zones of climatic spas	Health care zones
		Protection zones of spas and sanatoriums	
	Hospital grounds	Protection zones of hospital grounds	
Urban areas	Parks and public greenery, cemeteries	Sports, recreation areas, rest zones	Urban functional zones
	Sports areas		
	Individual housing development	Residential areas	
	Dwelling housing		
	Development of civic amenities	Pedestrian zones, zones of public services	
		Monument zones and reservations	Cultural heritage
		Cultural monument objects	
Communal-technical objects	Landfills	Hygienic and protection zones of landfill	Technical zones
	Ponds	Hygienic and protection zones of ponds	
	Incinerators	Hygienic and protection (security) zones of incinerators	
	Wastewater treatment plants	Hygienic and protection zones of wastewater treatment plants	
	Stores	Hygienic and protection zones of stores	

(continued)

Table 3.4c (continued)

Secondary landscape structure: elements of land cover		Tertiary landscape structure: socio-economic factors/phenomena	
Transport and telecommunication objects	Ports	Hygienic and protection zones of ports	Traffic and storage areas
	Airports	Hygienic and protection zones of airports	
		Safety zone of landing cones of airports	
	Railways	Protection belts of railways	
	Roads	Protection belts of roads	
	Power lines	Protection belts of power lines	
	Gas and oil pipelines	Protection belts of gas and oil pipelines	
	Telecommunication lines	Protection belts of telecommunication lines	

sive factors of natural differentiation in the landscape and fundamentally affects the horizontal and vertical flow of energy materials in the landscape. The interpretation of georelief provides knowledge on the dynamics of several current and past processes in the landscape. The significance of relief for the role of most reliable leading element (in addition to land cover) and the role of relief as a differentiation factor of landscape-ecological processes is discussed by many works (Krcho 1973; Hrnčiarová and Miklós 1991; Midriak 1999; Miklós et al. 1984; Mosimann 1984, 1990; Mičian and Zatkalík 1990; Miklós 1991; Tremboš 1991, 1994; Minár and Tremboš 1994; Minár 1994, 1995, 1998; Minár et al. 2001 and others).

The theoretical foundations of these approaches have been elaborated by a number of domestic and foreign authors (e.g. Demek 1974; Kočická 2000, 2007, 2011; Leser 1991; Leser and Klink 1988; Mazúr et al. 1983; Mičian and Zatkalík 1990; Miklós et al. 1990; Mosimann 1984, 1990).

Syntheses do not merely mean mechanical superposition of maps; but representing the actual relations among the elements of an abiotic complex. Therefore, certain combinations of properties of the elements are absolutely out of the question, unrealistic or very unlikely. Nevertheless, if in the process of syntheses such combinations occur, there has certainly been an error, either in the analytical maps or in the process of syntheses. These combinations have to be reviewed and corrected!

Table 3.4d Relationship of elements of the secondary landscape structure to the tertiary landscape structure: SEF related to industry—agricultural objects—exposed substrate

Secondary landscape structure: elements of land cover		Tertiary landscape structure: socio-economic factors/phenomena	
Industrial and mining objects	Objects and areas related to mineral resources	Natural rock structures for gas storage	Industrial and mining zones
		Exclusive mineral deposits	
		Protected deposit area	
		Mining space	
		Protection and security zones of mining areas	
		Declared undermined territories	
		Protection and security zones of stone-pits	
	Heaps, sedimental and disposal ponds	Protection and security zones of heaps and sedimental ponds	
	Industrial objects and areas	Protection, security and hygienic zones of industrial facilities	
	Construction objects	Protected (fenced) construction sites	
Agricultural-technical objects	Agricultural—technical objects	Protected (fenced) areas of machineries	Agricultural and technical zones
	Feedstocks	Protected (fenced) areas of feedstocks	
	Livestock farms	Protection and hygienic zones of livestock farms	
	Dumps of fertilizers	Protection and hygienic zones of dumps of fertilizers	
	Objects of agrochemical enterprises	Protection and hygienic zones of agrochemical enterprises	
Military objects	Field airports	Protection zones of field airports	Military zones
	Military objects and areas	Security and guarded areas of military objects, shooting-gallery, etc.	
		Guarded training areas	
Exposed substrate	Clay pits, borrow pits	Protection zones of stone-pits, sandpits, gravelpits	
	Sand, gravel		
	Rocks and reefs		

Table 3.4e Relationship of elements of the secondary landscape structure to the tertiary landscape structure: SEF related to the deterioration

Secondary landscape structure: elements of land cover		Tertiary landscape structure: socio-economic factors/phenomena	
Landscape as a whole	All landscape elements	Declared territories with the most affected environment	Deterioration of environment
	Predominantly urbanized landscape	Declared sources and areas with air pollution at different levels (above the limit)	Threat of natural resources
		Areas with radiation, noise, vibration, odour	
		Registered sources of water pollution of different sizes	
		Segments of polluted water flows in different classes of pollution	
		Sensitive areas (water act)	
		Vulnerable areas (water act)	
	Predominantly agricultural landscape	Areas with contaminated soils	
		Areas with hardened soils	
		Areas with eroded soils	
		Areas with landslides	
	Predominantly forest landscape	Forests affected by imissions	
		Forests affected by pests	
	Whole landscape	Territories threatened by avalanches	
		Areas with seismicity	

The process is schematically shown in Fig. 3.1a, b:
The models of synthetic units are named as

geocomplexes

of various orders and complexity, where models' elements are the indicators of their properties $\boldsymbol{x_n}$ (state variables or spatial subsystems), generally as

$$G_K(x_1, x_2, \ldots x_n)$$

Sites with the same combination of analytical values arise repeatedly over a sufficiently large territory, therefore we can say that they are typological units—**types of complexes** (e.g. abiocomplexes, landscape-ecological complexes)

(b) **Choric—Spatial—Structural Synthesis—Formation of Regions of Complexes**

There are many different approaches to choric syntheses. The LANDEP method mostly uses syntheses aimed at creating hierarchically higher synthetic units—regions according to the characteristic spatial and content grouping of lower hierarchical units, in our case types of abiocomplexes (ABC) or landscape-ecological complexes (LEC). Within their borders, ABC and LEC have unchanging (homogeneous) values for all the selected indicators $\boldsymbol{x_n}$.

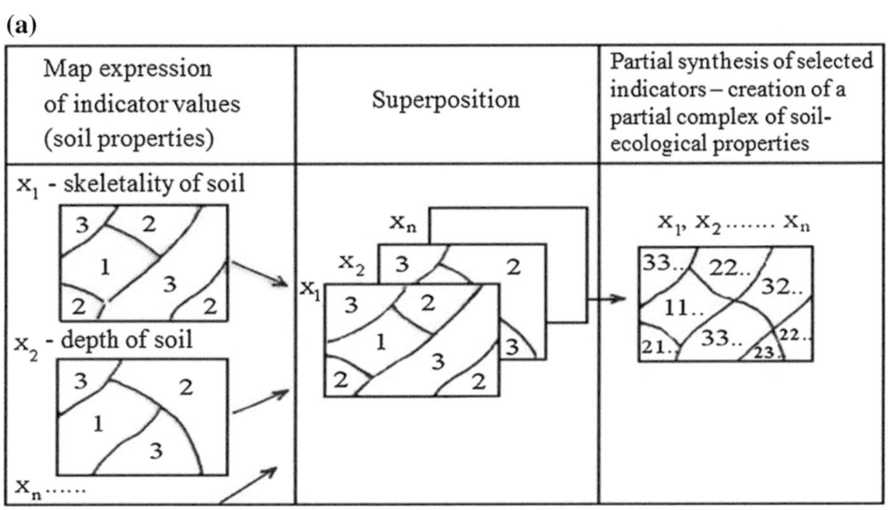

Fig. 3.1 **a**, **b** Schema of synthesis by superposition of analytical documentation

(b)

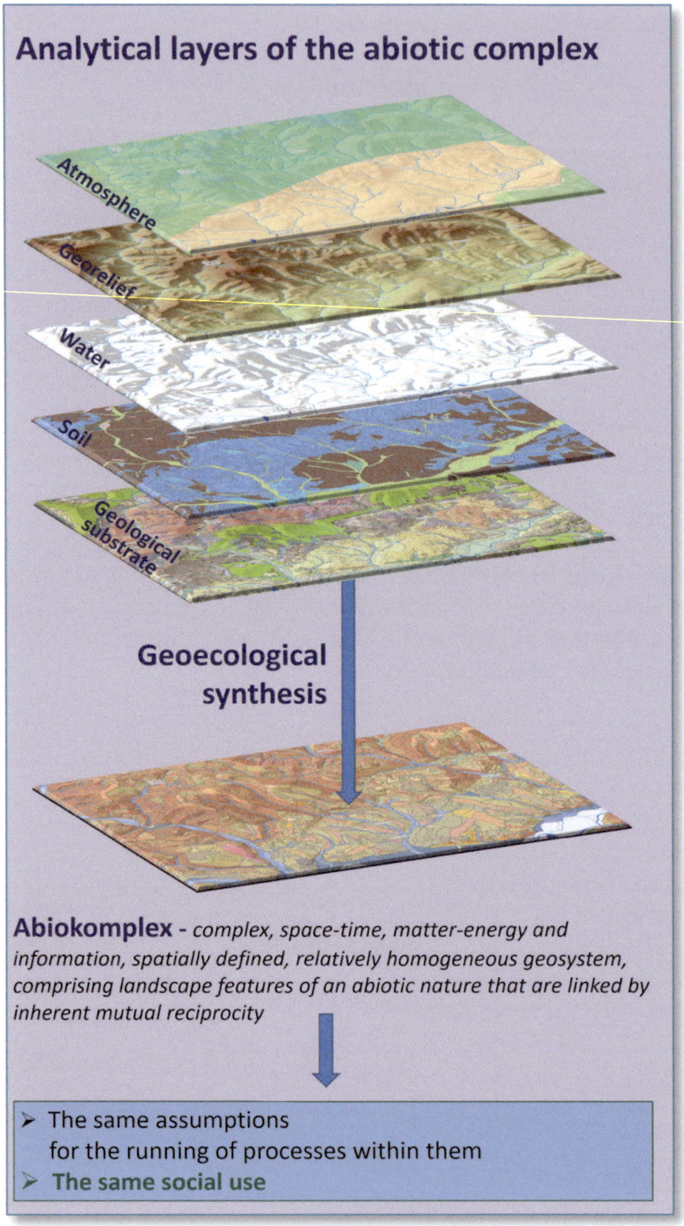

Fig. 3.1 (continued)

Homogeneous sites with exactly defined values of the indicators of the types of ABC or LEC constitute the **spatial database** for the next steps of different researches or applied methods, as the LANDEP method. A set of values of the indicators of analytical properties of a given LEC type represents certain suitability for its use by man in its entire homogeneous area and all other areas of a given type, i.e. it is a set of landscape-ecological conditions for deciding on the optimal utilization of the landscape (Mander et al. 2005; Szabó et al. 2008; Špinerová 2010; Miklós and Špinerová 2011, 2013). The synthetic units thus become basic **operational units** of the subsequent decision-making process.

Synthesis is a key step in the creation of geosystems. It can be said that it is also the final step in the basic research on the landscape, as well as the initial step in the applied procedures and planning practice.

3.2 Types of Geocomplexes

Types and regions can be regarded as a spatial framework, the material basis and carrier of information about the landscape. Therefore, by typification and regionalisation of the landscape we simultaneously create a **spatial database** where the georeference base is delimited sites of types and regions of the landscape with attributes of landscape features these sites characterize or by which we define them. These procedures and their results have scientific, cognitive and educational significance, as well as practical significance.

The practical importance of typification and regionalization of the landscape lies in the fact that for every created typological or regional unit it is possible to develop a full set of special characteristics, interpretations and assessments, regardless of their particular place of occurrence. It is also possible to draw up proposals for their optimum utilization and management measures. As the types and regions are a spatial database of accurately identified and defined sites, such interpretations of the evaluation, management measures can be transferred to these sites very simply (Miklós and Izakovičová 1997; Miklós and Špinerová 2011).

In recent times, this approach has been facilitated by the spatial database being made in the GIS environment, allowing an unlimited number of extremely varied interpretations and proposals. Typification of the landscape in this regard is therefore a key step in many applied landscape-ecological methodologies focused mainly on planning and spatial organization of the landscape (Malík et al. 2011; Diviaková 2011; Miklós et al. 2014; Kočická and Kočický 2014).

According to the above theoretical principles in a variety of specific papers, complex spatial units are created that can pragmatically be considered objective-oriented geocomplexes. Objective-oriented geocomplexes, for the needs of integrated landscape management will be of varying content and order. According to the internal contents, the following objective-oriented geocomplexes are routinely produced:

3.2.1 Abiotic Complex

An abiotic complex (abiocomplex, ABC) is a spatial unit synthesizing topical abiotic features of the primary structure of the landscape.

ABC is generally understood as complex, space-time, matter-energy and information, spatially defined, relatively homogeneous geosystem, comprising landscape features of an abiotic nature that are linked by inherent mutual reciprocity. Such an entity creates living conditions for plants, animals, human society, its socio-economic activities and products with abiocomplexes enter into intensive interaction.

Abiocomplexes are inherently quasi-homogeneous spatial units in terms of all considered properties of elements of the primary landscape structure, as well as the processes and phenomena taking place within the borders of these units. The homogeneity of complex units of the same type gives the same conditions for the conduct of the processes within them and thus equal conditions for use by society, making these units the most widely used information spatial database for most applied landscape-ecological research.

The simplified pragmatic recording of abiocomplexes is as follows:

$$\mathbf{ABC} = (\mathbf{Re}, \mathbf{Ge}, \mathbf{So}, \mathbf{Wa}, \mathbf{Cl})$$

wherein the elements of the complex are relief (symbol Re), rocks—geological base and substrate (Ge), soils (So), water (Wa) and air (Cl).

3.2.2 Physical-Geographical Complex

It is defined by a set of selected topical characteristics of the primary landscape structure, as defined in the ABC, with the explicit indication of the potential natural vegetation (**PNV**).

We emphasize that the PNV is notional construction; it does not exist itself (there is only real vegetation). PNV is a purposed interpretation of the ABC and understanding of the ecological conditions of occurrence of real vegetation. Nevertheless, it can be a considered complex interpretation of the characteristics of the primary landscape structure and PGC for potential geo-ecosystems (Miklós, Izakovičová et al. 2005, 2006).

The PGC model can be recorded as

$$\mathbf{PGC} = (\mathbf{ABC}, \mathbf{PNV})$$

3.2.3 Complex of the Current Landscape Structure: Land Cover and Real Vegetation/Habitats

Elements of the secondary/current landscape structure **(CLS elements)** are made of material elements that cover the earth's surface at present and that mankind has affected, remodelled or created. They are characterized to varying degrees of detail, as well as by a variety of functional aspects, such as elements of land cover (the general aspect), as elements–forms—of land use (land use, economic and geographical aspect), according to their bio-ecological content as physiognomic-ecological formation of real vegetation, as well as habitat (biotic aspect). From a physical point of view, they have the character of physiognomic-ecological objects. We regard them as synthetic space units—**CLS complexes**, in this understanding we consider them as the mapping units of CLS.

It is important to characterize the relationship between the terms land cover, land use, element CLS element, habitat. Guth and Kučera (1997) distinguish land cover as a real (physiognomic) situation in the landscape and land use as a listed type of land or culture. A more suitable resolution is provided by Feranec et al. (1996). According to these authors morphostructures and physiognomic characteristics generally correspond to the essential functional features to indicate the spatial organization of the cultural landscape, the land cover map presents the objects of the Earth's surface through its physiognomic attributes, the map of land use through functional attributes. From the geosystem perspective, however, it is still the same material (physiognomic-ecological) element, characterized from different perspectives. Within the LANDEP methodology therefore we use the umbrella term element of CLS, which is characterized as a CLS mapping unit.

$$\mathbf{CLS} = (\mathbf{INF}, \mathbf{REF}, \mathbf{CRE}),$$

where the elements of the complex are elements of land cover with are influenced (INF), reformed (REF) and created (CRE) by mankind.

If was give specific emphasis to the biotic complex, as well as the anthropic properties of the elements of CLS, such units may be regarded as the **biotic-anthropic complex (BAC)**. It is a defined set of elements of land use with real vegetation. In terms of material and space it is identical to the CLS complex, differences may be the level of detail the characteristics of their contents.

3.2.4 Landscape-Ecological Complex

ABC and CLS or BAC form the **landscape-ecological complex (LEC)**. LEC is defined by a complex of elements of the primary landscape structure (ABC) and the elements of the CLS, or BAC.

$$\textbf{LEC} = (\textbf{ABC}/\textbf{PNV}, \textbf{CLS}/\textbf{BAC}) \text{ or}$$
$$\textbf{LEC} = (\textbf{PGC}, \textbf{CLS})$$

3.2.5 Socio-economic Complex

Is a set of **intangible elements, factors (phenomena)** of the nature of **interests, manifestations and consequences** of human activities in the landscape that form the **tertiary landscape structure**, i.e. a **set of socio-economic factors/phenomena (SEF)**. The character of SEF applies to conservation of nature and natural resources, as well as negative character of SEF stress factors and other events.

Specific SEF mapping units have expression as

- boundaries of declared nature conservation areas;
- protected natural resources areas, zones of hygienic protection of water resources;
- buffer zones and safety zones of production, transportation and other technical facilities;
- administrative and sectorial boundaries;
- boundaries of sites, areas, sections of declared deteriorated territories of the environment.

The socio-economic complex SEC is formally defined as the complex of different groups of SEF:

$$\textbf{SEC} = (\textbf{NAC}, \textbf{NAR}, \textbf{URB}, \textbf{CAT}, \textbf{IND}, \textbf{AGR}, \textbf{ADM}, \textbf{DET}),$$

where: NAC—SEF of nature conservation, NAR—SEF of protection of natural resources, URB—SEF of urban, residential and recreation areas, CAT—SEF of communication and transport, IND—SEF of industrial and mining activities, AGR—SEF of agricultural activities, ADM—SEF of territorial units and administrative boundaries, DET—SEF of deterioration of natural resources and the environment.

A large part of the mapped SEF binds to the material elements of the current landscape structure. These, apart from being mapped out within CLS, are seen as SEF of primary stress factors.

3.2.6 Integrated Geocomplex

In the theoretical understanding, integrated geocomplexes IGEC are the resulting spatial units, integrating all spatial data on the territory. IGEC thus consist of a set of topical abiotic characteristics of the primary landscape structure—ABC, information on potential natural vegetation—PNV, information on land use and habitats—BAC and information on socio-economic factors/phenomena—SEC. Formally we can define the IGEC structure as follows:

$$\mathbf{IGEC} = \{(\mathbf{ABC},\ \mathbf{PNV}), (\mathbf{CLS},\ \mathbf{BAC}), \mathbf{SEC}\},$$

or more simply

$$\mathbf{IGEC} = (\mathbf{PGC},\ \mathbf{CLS},\ \mathbf{SEF})$$

It must be said, however, that in practice the model territory is characterized according to the objectives of the work. This means that usually some complexes are characterized in more detail, others in less detail, or not even addressed. In this sense, we consider IGEC as only a theoretical construct. In practice, it is almost never used. The most used concept in practice is the characteristic of landscape-ecological complexes LEC constituted by abiocomplexes ABC and current landscape structure CLS, completed by a complementary set of information on socio-economic factors SEE, i.e.

$$\mathbf{LEC} = (\mathbf{ABC},\ \mathbf{CLS}), \mathbf{SEC}.$$

To the above-described objective-oriented geocomplexes we can assign any additional characteristic, they can be interpreted, evaluated, applied in different ways. This feature makes geosystems an appropriate spatial and content framework and basis for any landscape-ecological assessments in sectoral planning procedures representing various institutional instruments of ILM. Such geocomplexes can be used at all hierarchy levels of ILM, from the national level down to the local level.

3.3 Integrated Spatial Information System—An Imperative Necessity for Researchers

As the spatial information system named as geographical information systems GIS is based on the geosystem theory, many authors have projected it into geography (Chorley and Kennedy 1971; Sochava 1978; Krcho 1991; Miklós and Izakovičová 1997; Kocsis et al. 2016). An integrated cartographic information system enables the comprehensive study also of very complex spatial and functional relationships. The GIS enables to study the landscape in 4D, which means the study of the land with changing time.

The fundamental aspect of such a database from the GIS perspective is the correct definition of the structure of the geosystem in the sense of the chain of terms **elements (component)—properties—indicators—values**. To achieve any result, the functions of these must not be confused when using the GIS technology. Accordingly, actual quantitative and qualitative characteristics of the landscape at each site of the landscape are than indicated by the specific values of indicators of properties of elements of the geosystem (Miklós et al. 2011a, b). Such a database can provide an integrated set of processed information practically for all planning processes.

The integrated information system is prescribed by the INSPIRE Directive 2007/2/EC (INSPIRE—Infrastructure for Spatial Information in Europe). It obliges EU Member States to operate a spatial information system on a united cartographic basis, with collection and processing of unified information (e.g. catalogue of objects) based on GIS technologies. The principles of the INSPIRE Directive are incorporated into Slovak legislation by Act 3/2010 Coll. on the **national infrastructure for spatial information**—NISI. This, inter alia, in §1 prescribes the '*establishment of a national infrastructure for spatial information for management and decision-making in public administration; coordinating the development and operation of the national infrastructure for spatial information; establishment of a national geo-portal*'. This system actually also covers the system of landscape information.

The cited act further states: '*The national infrastructure for spatial information means metadata, spatial data, sets of spatial data, services of spatial data, network services and network technologies, agreements on sharing, access and use of data for any activity for which this data is useful, coordination and monitoring mechanisms, processes and procedures, established, operated or made available; Metadata is information describing spatial data and spatial data services that enables them to search, inventory and use it; spatial data is an identifiable collection of spatial data*' (§2). Such a system therefore must contain processed data on primary, secondary and tertiary landscape structure using the (geo)system approach, and their expression in the form of spatial complexes. The factual content for this system may be the existing database, which under the name Landscape-Ecological Basis for Integrated Landscape Management (LEB ILM) has been provided by the Ministry of Environment of Slovak Republic (MoE SR) since 2000. The goals of LEB ILM have been formulated as follows:

- create **multi-functional usable documentation** for all programme objectives of the MoE SR (analytical databases, abiocomplexes, landscape-ecological complexes) as:
- apply documentation for objective solution as part of integrated landscape management, in particular integrated river basin management plans,
- create documentation of representative geo-ecosystems;
- create documentation for integrated protection and management of protected areas, including identification sheets of protected areas, revision of borders and degree of protection;
- create documentation for spatial planning—landscape-ecological planning, ecological networks;

Creating and using LEB ILM has been directly prescribed by Act 7/2010 Coll. on flood protection, which defines LEB ILM in Sect. 3.5, Preliminary Flood Risk Evaluation, as follows: '*The basis for the preliminary flood risk assessment and flood risk management plans also includes the evaluation of run-off conditions, the conditions for the emergence of floods as well as the conditions for increasing the retention capacity of the landscape. As a base there will be developed spatial data of on relevant elements of abiotic complexes in the river basins, particularly morphometric characteristics of the relief, the physical properties of soils and geology, as well as spatial data on the elements of the current land-use. The collection, processing, storing and providing relevant spatial data will be arranged by MoE SR under the National Infrastructure of Spatial Information—NISI*'. The current version of the act has dropped the term LEB ILM and the act in Sect. 3.5 states: '*The preliminary flood risk assessment shall be drawn up on the basis of information that is available or can be easily obtained …*'. The fact remains that the frame and the framework for all sectoral planning procedures should be a physical (territorial, spatial) plan created on landscape-ecological foundations. In the Slovak Republic this is provided by the above-mentioned methods of landscape-ecological planning (LANDEP) and territorial system of ecological stability (TSES), which forms the legal basis for the process of physical/territorial planning (Act 50/1976 Coll., on territorial planning and the building code—Building Act—as amended, in particular by Act 237/2000 Coll.).

The united cartographic basis for Slovakia lies on a frame of geodetic and cartographic information called as Fundamental Base for Geographic Information Systems (ZB GIS, Koncepcia … 2006; Collective 2008). The system is supplemented by factual information collected within GEOFOND (the state fond of geological information in Slovakia operated by the State Geological Institute of Dionýz Štúr), as well as by various atlas works, catalogues and other spatial data from different sectors. The aim of these instruments is to ensure a united information basis for all sectoral planning procedures.

The united information system of geocomplexes represents

- a comprehensive system of information on the landscape, an object-attribute spatial database;
- they are the base for creating the interpreted purpose-oriented characteristics of the landscape;
- the object of evaluation for selected activities of society;
- they create the boundaries of areas for proposal of optimal use of the landscape, i.e. they are the subjects of proposals.

3.3.1 Methodological-Practical Problems—The Use of GIS in Solving Analytical and Complex Problems

As said, many 'classical' sciences cannot easily follow the advancement of the geographical information systems due to the fast development of computer science and information technology. This explains why the technical capacities of the GIS are not yet fully used. On the other hand, the GIS specialists have not answered all the problems that are considered elementary by the users from various sciences. The problems of using GIS mostly arise when we want to analyse the complex relationship of components in the geosystem and want to use an information system based on GIS.

The map is the basic tool to understand, scientifically describe and visualize the space. The maps will be the basis of geographical information systems even if they are not produced as conventional prints but presented in electronic form. The most important property of a map, that is the definition of the location of each point in space by a coordinate system [x, y, z], has remained the basis of the electronic GIS. The material objects of the geosystem are also represented in this coordinate system.

There are still several basic methodical questions, some of which follow bellow.

3.3.2 Geometrically Unified Projection System and Unified Topography—Unified System of the Surface and Objects

This is not a problem in itself for information technologists, because it is expected from every good atlas that the maps are prepared in the same scale, and they have the same projection and execution. Organization problems, however, do exist. For instance, it was n=ot, and is not always possible to expect from the authorities that they place all the necessary information on the same cartographic base.

The unified topography is not a problem in itself for information technologists either. Modelling the surface is not a difficult job at all, because this is done by the computer by using the digital model of the terrain. However, harmonizing the material components existing in space, the basic (topographic) map elements on the one hand, and the elements of thematic maps on the other hand, has been a problem that has to be solved urgently.

The basic topographic map elements often appear in different positions and locations in maps of various origins even though they were prepared in the same projection. It may happen that a stream electronically transferred from a map onto another map flows on the hillside and not in the valley, the roads are sometimes entwined like someone's hair in plaits, rendzina soils come over the alluvial deposits, acid forest soils cover the limestone slopes, the lake is placed on a ten-degree slope, etc. These mistakes lead to chaotic combinations in the complex assessment and in overlapping, and they may produce misleading results.

The experts of informatics state that they can 'transform' any map projection into a unified basis. However, this is only the transformation of the coordinate system! The original thematic information (and not the multi-scanned copies) of thematic maps was manually placed on the map, and this content cannot be transformed by mathematical rules. If we need the complexity of information and we want to analyse the interaction of individual components, we have to standardize them and we have to exclude the absurd combinations. The basic elements of the topographic map and the basic elements of land use can be accurately projected into their uniform position only manually and with expertise as well as with the comprehensive understanding of the relationships between the components of the geosystem. The availability of orthophoto maps and georeferenced space images, which can be handled electronically, is of great help today. In fact, it is the discipline of the complexes of abiotic, social-economic or other spatial units which should determine the quality of the information system.

In spite of problems, there are developments in this direction too. In Slovakia, for instance, the Geodesy, Cartography and Cadastre Authority of the Slovak Republic, the Institute of Geodesy and Cartography, and the Topographic Institute (formerly the Military Topographic Institute) worked out together a GIS-based catalogue of objects (Koncepcia … 2006, Katalóg 2008). This catalogue has been continuously harmonized with the Central Regional Database of the Military Regional Information System, updated with the georeferenced aerial photographs (orthophoto maps), and uniformly presented in maps. During the execution of the GIS project on the Ipoly drainage basin, we used this information basis.

The Institute of Landscape Ecology of the Slovak Academy of Sciences in Bratislava worked with 'manually made' GIS systems. The computers were substituted by tracing paper and transparent foils, and the foil layers were placed onto each other to gain complex information of interactions necessary for the analyses (Kozová et al. 2007). The projects included landscape-ecological plans developed by LANDEP for various levels, from country level (known as Ecological Generel) to quite small areas like the Ipoly drainage basin. The largest scale project of these 'manually made GIS-based projects' was the landscape-ecological plan prepared for the East Slovakian Plain at 1: 25,000, where more than fifty thematic map layers were harmonized on a uniform cartographic basis (Miklós et al. 1986a). However, computers were already also used towards the end of the project. This was of great methodological importance, because it showed what a computer can do and cannot do (Miklós et al. 1986b). There are three new atlases made by the Institute of Landscape Ecology worth mentioning, which were completely prepared by using GIS technology: Landscape Atlas of Slovakia (Miklós and Hrnčiarová et al. 2002), Atlas of the representative geo-ecosystems of Slovakia (Miklós et al. 2006), Atlas of the geo-ecosystems of the Trnava district (Izakovičová et al. 2011). The experiences gained during their preparation were implemented in the development of the GIS for the Ipoly drainage basin (Verrasztó 2010; Miklós et al. 2011a, b).

The difficult task of landscape ecologists regarding the projection, representation, raster–vector transformation is to solve the following questions:

- how to structure the information into data components for the planned thematic maps so that the relationships between the impact sources and affected factors can be analysed,
- which projection and representation method should be used for the individual thematic layers so that they can be transferred from vector format into raster format or vice versa, and the thematic layers can be combined and properly interpreted.

These phases also make great use of the experiences collected during the time of the manually building of the GIS.

3.3.3 The Elements of Georeferencing

Although this point seems to be a trivial task for IT professionals, this is of great significance concerning the usability, filling up and updating—that is the management of the future monitoring—of the system.

The elements of georeferencing can be divided into two groups:

(a) Elements of Georeferencing for the Determination of Place and Position—Primary Spatial Location

All points of the surface of the Earth are described by the geographical coordinate system (that is by geographical latitude, longitude and elevation above sea level—φ, λ, h) or by a geometric coordinate system (x, y, z coordinates). This coordinate system makes it possible to create the digital terrain model (DTM) of the surface. This is in fact a coordinate system of the intersection points of the grids of predefined side length (say of 10 × 10 m). Calculating the morphometric indices of the surface and their representations are based on the DTM, which has been of immense importance in landscape-ecological research. The DTM is also the basis of forming various isolines.

Naturally, it is the geographical or the geometrical coordinate system that determines the place and position of all the other elements of georeferencing. They are considered secondary elements of georeferencing.

We should not forget that, in addition to those mentioned above, recording the time factor in the system is also of great significance, because the environmental changes can be studied only on condition the past states of the environment are processed in the same map system. The proper cartographic combination of the impact sources and affected factors makes environmental modelling possible.

(b) Elements of Georeferencing that Carry the Properties of the Material Components of the Geosystem

Naturally, the grid of the coordinate system mentioned above can be used to locate any component in space. It can be mainly used for the representation of those material components that quickly change in time and for which it is impossible to create a constant georeferencing element (e.g. the momentary location of a thundercloud, the spread of air pollution). However, the following elements of georeferencing are used much more frequently and effectively for the material components.

- raster—this is a grid with regular sides too. However, the georeferencing does not deal with the intersection points, but it refers to the area of the mesh. This gives us the great opportunity of representing, analysing and comparing numerous components of the geosystem. Therefore, it can be used in many ways;
- vectors—they define components as points, sections and polygons.

These are the carriers of the geographical information in the most important topographic and thematic maps. These components can be used in lots of areas and can be excellently used to fill up a system with data continuously. Let us mention that the complex spatial units are also georeferenced according to polygons. The indices related to points give us the opportunity of compiling isolines.

Finally, the following can be stated on the elements of georeferencing:

- the properly chosen elements of georeferencing—that is the spatial carriers of the database—make it possible to update, reload and complete the information almost without any limitation as well as to compare all kinds of data at present and in the future without changing the spatial system;
- if the elements of georeferencing are not chosen properly, the whole data line may be used only once, and they cannot be combined or updated; a non-uniform system, though may be full of data, does not offer the logical combination of information therein.

Let us see a simple example. If the meteorological stations are the elements of georeferencing, they can receive newer and newer sets of data at any time independently of the indices we have at present. However, if we enter the properties of some atmospheric phenomenon into the system in the form of isolines only, they cannot be updated or completed even if they are in a large number or in excellent resolution. The new isolines will have to be digitized, georeferenced and represented.

The most frequent applications of the elements of georeferencing are shown in Fig. 3.2.

Representation of the elements of georeferencing	Name	Most frequent indices
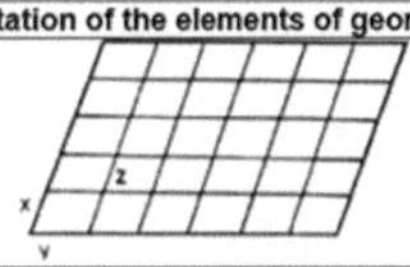	Grid of coordinates x, y, z φ, λ, h Pixel	DTM Morphometric indices Surface
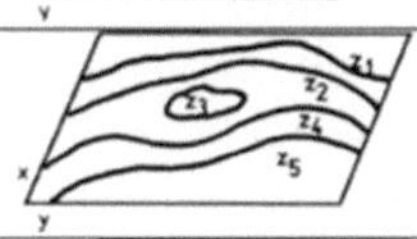	Isolines	Altitude (a.s.l.) Level of groundwater
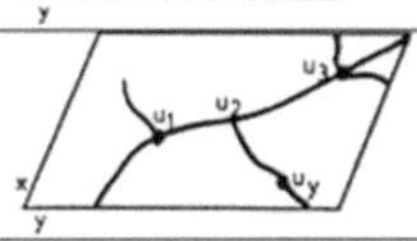	Point Section	Hydro and meteo stations River and road sections ID points of settlements
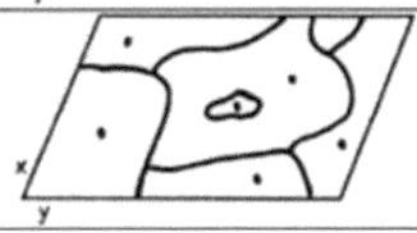	Polygon Reference points of polygon	Territorial units Complexes

Fig. 3.2 Most frequent applications of the elements of georeferencing

3.3.4 Content of the System—Indicators

The essence of information systems is the indices and values of the properties of the elements of the geosystem. The elements of georeferencing are their spatial representatives. It would be ideal to enter a large number of usable indices into the system, but there are some practical barriers. This explains why it is so important to choose the proper indices. Another important condition was already mentioned: this is the system of the proper elements of georeferencing, which, at a later stage, makes it possible to systemize the not yet classified indices, though they are already related to certain elements of georeferencing.

Naturally, the indicators are primarily chosen according to the aims of the information system. The main aim at present is collecting and systemizing information, and later it will be scientific research, planning and decision-making. The ideal state would be a multipurpose use of the system, which would be a system filled with lots of analytical information and completed with a filter/search module; in this case, the necessary information could be filtered for any purpose. At present, however, rather the minimal aims have to be defined and the system has to be built user-friendly.

When building the system, it is very important to know the degree of originality of the indices, in other words, to what extent these indices had been interpreted. Naturally, a good information system predominantly should contain elementary, analytical—that is primary—information, which allows the users to obtain derived and interpreted indices. The thesis is that if the analytical indices are correct, lots of derived data can be obtained and assessed! However, this idea does not work backward: even if the derived index is a very good one, maybe it is impossible to know the basic index! Let us see a simple example. If we know the usability of a soil or the degree of its erosion, we cannot deduce properties like the size of the soil particles or the angle of slope. However, if we know the angle of slope and the size of the soil particles, we can calculate the degree of erosion and the usability of the soil from these data; in addition, several other indices can also be interpreted.

These ideas are not new: landscape ecology and earth science always made a difference between causes and effects. This consideration got special emphasis in the study of environmental issues (e.g. Verrasztó 1979). These theses were first directly used in developing the LANDEP methodology for building a spatial information system and were described in the steps 'Analyses—Syntheses—Interpretations—Evaluations—Propositions' (Ružička and Miklós 1990) and successfully also in later projects.

3.3.5 Monitoring and Its Information System

Though monitoring has become a fashion word today and it has been scientifically justified, the term needs some explanation. In science, monitoring usually means continuous observation with the aim of collecting data in a series of time so that scientists can draw conclusions from certain processes, relationships and interactions. As for spatial information systems, the term, 'continuous' is of special importance, because the spatial elements of the geosystem can be monitored in various ways:

- continuously—without any break;
- constantly, but with breaks—periodically, in shorter or longer regular periods;
- seasonally, in various periods according to need.

Naturally, some of the elements or factors may have to be monitored in various periods: e.g., let us compare the monitoring of geology, land use or weather. It can be stated that monitoring rather means regularly repeated or renewed research in many cases.

The major question here is how the observed data, including the actual monitoring data, can be built into the spatial information system. It is important to note that the information system does not register the processes, neither the interactions nor the relationships, but 'only' registers the various states—preferably in different periods. However, from these states, scientists can already interpret the interactions and relationships. For this purpose, it is an important condition that the registered states of elements should describe the same location and preferably should refer to the same time or period. In other words, for the interpretation of observations the information system of monitoring is most important, namely:

- the system of georeferential elements, and
- their primary data.

Then follow further steps, which already may have the character of monitoring such as

- the continuous collection and storing of information;
- updating of time series, and
- distributing and using information.

3.4 Creation of the Database of Geocomplexes in Slovak Republic

The synthetic well defined and precisely mapped geocomplexes are considered the **basic operational units** of the decision-making process for integrated landscape management. For this reason, it should be the interest of each state to process such a spatial database for its entire territory, which would be unified and accessible to a broad layer of users.

Such a procedure is prescribed by the INSPIRE Directive 2007/2/EC, as well as Act 3/2010 Coll. on the National Infrastructure for Spatial Information (NISI). The Ministry of the Environment of the Slovak Republic (MoE SR) decided to build such an information base in terms of geosystem theory already in 2000, work was finished by processing completed abiocomplexes.

The first step for creation of the database on geocomplexes is the systematic and successive definition of all

- the elements of landscape structure, which should be included in the database;
- the properties of these elements, which are of interest for objective-oriented studies;
- the indicators/attributes of these properties;
- the characteristic values of the indicators of properties;
- the amplitude of the values.

For the whole territory of Slovak Republic a united spatial database of abiotic complexes was elaborated based on the theoretical principles stated above. The basic aspects of collecting and homogenizing the input data (landscape-environmental analysis and partial synthesis) and the sources for data can be summarized as follows:

- the spatial database is prepared on the basis of the basic topographical maps, S-JTSK_Krovak_East_North on a scale of 1:10,000, which is currently the most accurate nationwide cartographic basis. The fundamental basis for the creation the geological database was the digital geological map of Slovakia 1:50,000 (Káčer et al. 2005, updated). As part of the database a digital map of genetic types and thicknesses of quaternary was created on a scale of 1:50,000 and a digital map of soil types at 1:50,000 (Malík et al. 2007);
- creating the methodology, setting the content and specific delimitation of geocomplexes within the nationwide database has been heavily dependent on the basic data on relief, rocks and soil (maps of morphotopes, lithostratigraphic units, quaternary, soil classes and soil types. Information about other abiotic landscape elements (water bodies and air) has been added based on the interpretation of the relief-rock-soil complex and internal materials of bodies processing data;
- an important basis for creating maps of abiocomplexes was the digital map of geomorphological units at scale 1:10,000 prepared on the basis of the map of geomorphological division in the Landscape Atlas of the Slovak Republic (Miklós and Hrnčiarová et al. 2002), which is an adapted map of original by Mazúr and Lukniš (1986), map of geomorphological conditions of SR at scale 1:500,000 (Mazúr et al. 1980, adjusted in Miklós and Hrnčiarová et al. 2002), the map Morphological-morphometric types of relief (Tremboš and Minár In Miklós and Hrnčiarová et al. 2002);
- processing morphometric relief characteristics was supported by several theoretical works dedicated to this issue (Krcho 1973, 1983, 1991; Miklós et al. 1990; Miklós et al. 1991; Hofierka and Šúri 1996; Hofierka 1997 and so on.). Geomorphological regionalization, delimitation of elementary forms of relief, the problem of scale and the hierarchical order of geomorphological forms, has been extensively addressed in the works of Minár (Minár 1992, 1994, 1998). For a delimitation of positional units of relief, the base map was a processed digital version of the map of the geomorphological division of Slovakia by Mazúr and Lukniš (1986) and an additional basis was the map of morphological and morphometric types of relief of the Slovak Republic by Tremboš and Minár In Miklós and Hrnčiarová et al. 2002;

- an important result of creating the database is the creation of the digital terrain model (DTM) with a resolution of 20 m (DTM20-EN). Processing documentation takes place in the environment of GIS ArcGIS 9.3 and in the S—JTSK coordinate system.
- for the purposes of the detailed spatial allocation of abiocomplexes was necessary a revision over the borders of geomorphological units on the map of geomorphological division at 1:500,000 (Mazúr and Lukniš 1986). From the digital terrain model were derived: morphometric characteristics, digital geomorphological map at 1:50,000, digital maps of the river network, maps of horizontal and vertical dissection. The digital map of the river network was further corrected on the base of satellite images;
- other synthetic maps at 1:500,000 in from the Landscape Atlas of the Slovak Republic were adjusted (types of abiotic complexes, landscape-complexes, Miklós, Kočická, Kočický In Miklós and Hrnčiarová et al. 2002 etc.;
- after this process of precision, the geomorphological units with revised boundaries were then considered as the spatial framework for other characteristics of abiocomplexes. Such complex characterization also serves the needs of subsequent regionalization.

The database has been processed since 2003 by Esprit, s.r.o. The abiocomplexes were prepared in the form of a digital map that covers the whole country. Banská Štiavnica. It is defined using an extensive set of primary and interpreted (relative) properties of the abiotic complex. Originally, the database was filled for the needs of applied landscape-ecological processes, mainly as part of the 'Comprehensive Programme of Erosion Control and Proposal of Measures to Increase the Retention Capacity of the Territory of Slovakia broken down by sub-basins—basins of the Hron, Ipeľ, Rimava, Slaná, Poprad-Dunajec, Bodva, Bodrog, Hornád' (Kočický et al. 2004–2007).

More systematic elaboration of the acquired spatial database took place within the geological project 'Compilation of geological maps on a scale of 1:50,000 for the needs of integrated land management' under the umbrella of the State Geological Institute of Dionýz Štúr (SGIDŠ), in cooperation with the Research Institute for Soil Science and Conservation (RISSC) in Bratislava. These tasks emerged from the plan of main legislative tasks of the Section for Geology and Natural Resources at the Ministry of the Environment of Slovak Republic, where one of the objectives was to develop concepts of engineering and hydro-geological maps on a scale of 1:50,000 for the needs of Integrated Flood Management (IFM).

The original database of the abiocomplexes was completed in 2007 as part of the landscape-ecological synthesis as the basic documentation for landscape-ecological evaluation (Kočický et al. 2004–2007).

After 2006, the concept changed slightly. The task was expanded to integrated land management. The comprehensive form of the database emerged in 2011 under another task named as 'Comprehensive geological information base for the needs of nature conservation and landscape management (GIB-GES)'. The project included reviewing the existing database of abiocomplexes and adding complementary characteristics (aspect, altitude, insolation, climate type), position and regional characteristics of horizontal and vertical dissection, revision over the boundaries of geomorphological units on a scale of 1:50,000, assigning positioning units for relief with the use of previously conducted geological work under previous tasks. The output of the project was to create a united database, understanding of regional characteristics of the morphological and geological segmentation of the territory, processing of data on groundwater levels and flow directions of groundwater. These results were incorporated in the form of interesting reporting applications for any selected area in the Geological Environment Information System (GeoIS), which is currently under construction at State Geological Institute of Dionýz Štúr (SGIDŠ. The database is accessible in the archives of the Geofond of SGIDŠ.

Processing the database within geographical information systems (GIS) enables the derivation of more efficiently interpreted indicators, or the revision, updating and supplementation of existing parameters of objects, the derivation of parameters for units of a higher order, creation of additional interpreted geological, hydrological and landscape-characteristics. The database has a well-defined set of attributes, so it is easy to implement it into other information systems, and thus make it available to the professional and general public. The database will be maintained and supplemented as required by institutions, research and practice, as well as the development of information technology.

When creating a database, it was assumed that the created documentation is to be the cornerstone of regional geo-ecological studies, applied environmental studies and may be a binding basis for several landscape-ecological methods. It was also assumed that its publication would enable significant streamlining and improvement of processing of relevant documents (ecological networks, spatial planning, land consolidation projects, forestry plans, environmental impact assessment—EIA, and strategic environment assessment of documents on the environment—SEA).

In the compilation of the database, an important step is the standardization of landscape documentation. In the process of creating a database of abiocomplexes there are also formulated exact procedures for obtaining and demarcating spatial-ecological units. An important factor is the use of the information on georelief as a unifying information base for the content of all other geosystem elements and also for the spatial extrapolation of data. The issue of extending these maps in Slovakia has thus far not attracted sufficient attention. Typical are particularly approaches of the simplest method for creating abiocomplexes by mechanically placing sectoral maps often from different original scales and of significantly varying quality (especially the spatial accuracy of information) in the GIS environment without any correction. Especially on large scales, such an action is difficult to accept. All later, often demanding and fair, operations with a set of basic data that is doubtful will lead to a dubious overall result.

Therefore, during the creation of the database, the **illogical combinations were reassessed** on the basis of knowledge of legitimate vertical links between the attributes with each other and their relationship to climate, hydrological and biogeographic factors. It is a challenging and fundamental step, which consists of the creation of real, correct, synthetic relief- rock-soil units. Such units cannot be estab-

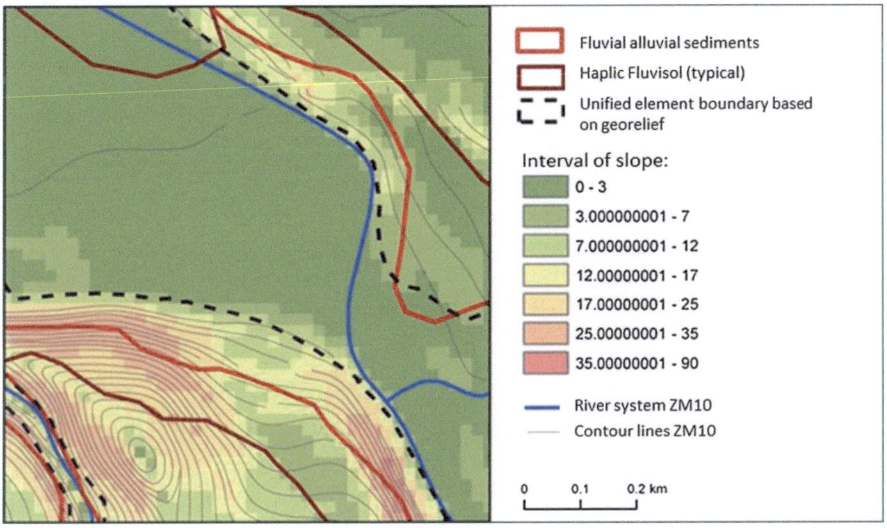

Fig. 3.3 Comparison of the course of boundary of geological substrate on geological map at 1:50,000 and boundaries of soil types on soil map at 1:50,000 and their comparison to the relief. Boundaries of depicted units of both elements should be the same and should be bound to morphographic type of alluvial plain (dotted line), which usually does not exceed 3° slope. Superimposing these layers without modification results in absurd combinations of parameters

lished merely by mechanical superposition, since the inaccuracy of the basic analytical documents causes illogical combination of parametric values of properties, small unreal areas, forms and positions of sites (Fig. 3.3). Synthetic units have therefore to be adjusted according to the logic of functional relationships within the abiotic complex and problematic areas verified in the field.

Each delimited site of these adjusted abiocomplex has a unique code made up of values of the indicators of the properties of elements of abiocomplexes—a selected set of values, which ensures logical consistency of attributes (Fig. 3.4). The result is a united spatial database covering the whole territory of Slovakia, serving for further action as a basic spatial database with such a set of abiotic characteristics that allow for each type of abiocomplexes to be evaluated for both the potential susceptibility to selected states and processes threatening the ecological quality of the landscape and on this basis, to draw up the proposal for optimum land use and location of ecostabilizing measures.

For the whole territory of Slovakia a detailed united spatial database of abiotic complexes was created. The current database of abiocomplexes is much more complex, with an extensive set of primary and derived data. In Slovakia over 925,000 sites of abiocomplexes have been identified.

The existing database of abiotic complexes have been effectively applied in several projects of regional and local territorial systems of ecological stability (RTSES, LTSES), in land consolidation projects, in the process of environmental impact

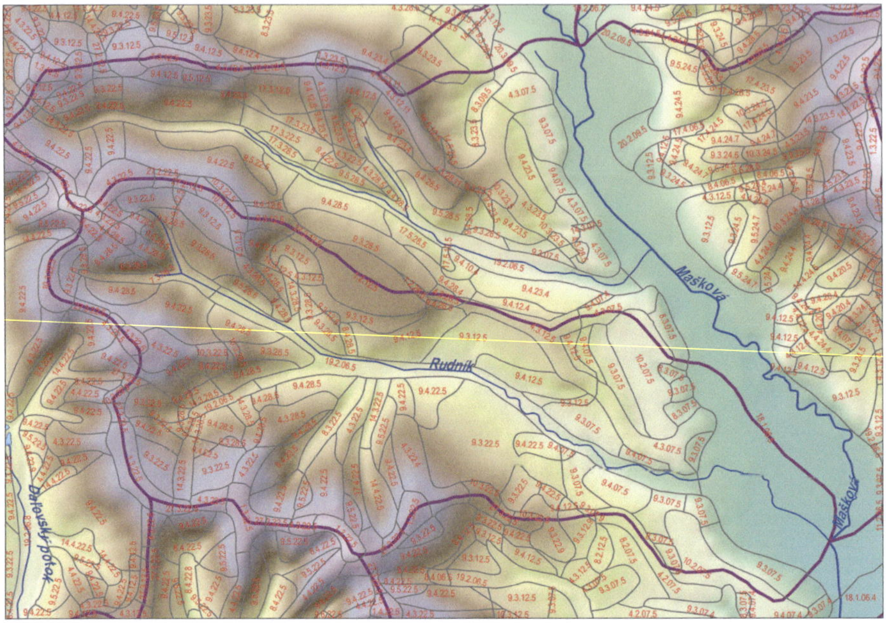

Fig. 3.4 A visualized illustration of a map of partial abiocomplexes of Slovakia (cut). The map is processed on original scale 1:10.000 in GIS. Each polygon displays homogenous area defined by values of four abiotic indicators (morphographic type, slope angle, geologic-substratum complex, soil texture). The violet line is the border of watershed

assessment. The database is also used in the conception of representative geo-ecosystems (REPGES). REPGES at the national level, processed on map scale of 1:500,000 (Miklós et al. 2002). In Slovakia, at 1:500,000, a total of 121 types of potential REPGES have been marked. These were the basis for their further processing. As a book, the concept was published under the title 'Atlas of representative potential geo-ecosystems'. Processing of maps of representative geo-ecosystems (REPGES) is part of the projects of local territorial systems of ecological stability. This processing is mainly supported by a digital database of abiotic complexes processed within the project 'Compilation of geological maps on a scale 1:50,000 for the needs of integrated land management' implemented by SGIDŠ from 2003 to 2007.

In addition to a database on abiotic complexes in Slovakia there are also nationwide databases available on potential natural vegetation, the current and tertiary structure of the landscape. These databases are built and managed by several institutions, much of the data is already published, much has also been made approachable online on the accessible servers. The creation of the database and its usefulness for landscape-environmental purposes is described in detail in the paper of Kočická and Kočický (2014). The mentioned database of abiocomplexes is based on the documented and credible analytical materials (Malík et al. 2007, 2011).

References

Chorley RJ, Kennedy BA (1971) Physical Geography—A System Approach. Prentice Hall Interantional Inc, London, p 370

Collective (2008) Katalóg tried objektov ZB GIS. Úrad geodézie, kartografie a katastra SR, Bratislava, Topografický ústav, Banská Bystrica. 229 pp

Demek J (1974) Systémová teórie a štúdium krajiny. Studia Geographica, 40, Brno, Geografický ústav ČSAV, 198 pp

Diviaková A (2011) Biotické komplexy pre environmentálny manažment. VKÚ, a. s., Harmanec, p 120

Feranec J, Oťaheľ J, Pravda J (1996) Krajinná pokrývka Slovenska identifikovaná metódou CORINE land cover. Geographia Slovaca, 11. Bratislava: Geografický ústav SAV, ISSN: 1210-3519

Guth J, Kučera T (1997) Monitorování změn krajinného pokryvu s využitím DPZ a GIS. Příroda 10:107–124

Hofierka J (1997) Modelovanie prírodných javov v prostredí GIS. /Dizertačná práca/, PRIF UK, Bratislava, 80 pp (nepubl.)

Hofierka J, Šúri M (1996) Modelling spatial and temporal changes of soil water erosion. Geografický časopis 48:255–269

Hrnčiarová T, Miklós L (1991) Morphometric indices in the interpretation of water and material motion dynamics illustrated on the example Dolná Malanta. Ecology (CSFR) 10(2):187–221

Izakovičová Z, Hrnčiarová T, Králik J, Liška M, Miklós L, Moyzeová M, Pauditšová E, Ružičková H, Šíbl J, Tremboš P (2000) Metodické pokyny na vypracovanie projektov regionálnych územných systémov ekologickej stability a miestnych územných systémov ekologickej stability. MŽP SR, Združenie KRAJINA 21, Bratislava, 155 pp

Izakovičová Z, Miklós L, Moyzeová M, Špilárová I, Kočický D, Halada Ľ, Gajdoš P, Špulerová J, Baránková Z, Štefunková D, Kenderessy P, Šatalová B, Dobrovodská M, Hrnčiarová T, David S, Krnáčová Z (2011) Model reprezentatívnych geoekosystémov na regionálnej úrovni (Model of representative ecosystems at regional level, in Slovak). Ústav krajinnej ekológie SAV, Bratislava, Slovakia, p 86

Káčer Š et al. (2005) Slovenská republika—digitálna geologická mapa v M 1:50,000 a 1:500,000. Manuskript—archív odboru Geofondu ŠGÚDŠ Bratislava, 42 pp

Katalóg tried objektov ZB GIS [Catalogue of object classes ZB GIS, in Slovakian] (2008) Úrad geodézie, kartografie a katastra SR, Topografický ústav. Bratislava, Banská Bystrica, 229 pp

Kočická E (2000) Optimisation of the territory exploitation with an example of the complex physical geographic evaluation. Ekológia (Bratislava), 19: 92-98, Supplement 2/2000 Kočická E (2001) Geoekologická (komplexná fyzicko-geografická) mapa ako podklad pre hodnotenie ekologickej únosnosti krajiny. ACTA FACULTATIS ECOLOGIAE (Zvolen) 8:31–38

Kočická E (2007) Stanovenie ekologickej únosnosti na základe citlivosti krajiny na acidifikáciu (modelové územie povodie Belianskeho potoka—Štiavnické vrchy)./Monografia/, Vydavateľstvo Technickej univerzity vo Zvolene, 135 pp

Kočická E (2011) Príklad ohraničovania komplexných priestorových jednotiek pre krajinno-ekologické hodnotenia. Acta Facultatis Ecologiae (Zvolen), TU Zvolen, 24–25:55–65, ISSN 1336-300X

Kočická E, Kočický D (2014) Databáza abiotických komplexov ako krajinno-ekologický podklad pre integrovaný manažment krajiny v Slovenskej republike. Acta Facultatis Ecologiae (Zvolen), TU Zvolen, 31:35–50, ISSN: 1336-300X

Kočický D, Maďar D, Maretta M, Hronček S, Foltán M, Švec P, Ivanić B, Kočická E, Kovačovičová Z, Špilárová I, Szolgay J, Hlavčová K, Kohnová S, Parajka J, Skalová J, Kubeš R, Zvolenský M, Jaroš B, Danáčová M, Svetlíková D, Tegelhoffová M, Danihilík R, Papánková Z, Horvát O, Gaál L(2004–2007) Komplexný program protieróznej ochrany a návrh opatrení na zvýšenie retenčnej schopnosti územia SR v členení podľa čiastkových povodí—povodia tokov Hron, Ipeľ, Rimava, Slaná, Poprad-Dunajec, Bodva, Bodrog, Hornád. Záverečné správy + prílohy, Banská Štiavnica: Esprit s.r.o

Kocsis K, Agárdi N, Koczó F, Mezei G A, Nemerkényi Zs, Szabó R (2016) Generalizálás géppel, vagy kézzel?—módszertani összehasonlítás Magyarország Nemzeti Atlasza példáján. In: Pajtókné Tari I, Tóth A (eds) Magyar Földrajzi Napok: Absztraktkötet : VIII. Magyar Földrajzi Konferencia. Budapest: Magyar Földrajzi Társaság, p 103
Koncepcia tvorby, aktualizácie a správy ZB GIS na roky 2006–2010 (2006) Úrad geodézie, kartografie a katastra SR, Bratislava
Kozová M, Hrnčiarová T, Drdoš et al (2007) Landscape Ecology in Slovakia. Development, Current State, and Perspectives. Monograph. Contribution of the Slovak Landscape Ecologists to the IALE World Congress 2007 and to the 25th Anniversary of IALE. Bratislava: Ministry of the Environment of the Slovak Republic, Slovak Association for Landscape Ecology—IALE-SK, 2007, CD ROM, 541 pp
Krcho J (1973) Morphometric analysis of the relief on the basis of geometrical aspect of field theory. Acta Geographica Universitas Comenianae (Bratislava), Bratislava 1:3–233
Krcho J (1983) Teoretická koncepcia a interdisciplinárne aplikácie komplexného digitálneho modelu reliéfu pri modelovaní dvojdimenzionálnych polí. Geografický Časopis, Bratislava 35:265–291
Krcho J (1991) Georelief as a subsystem of landscape and the influence of morphometric parameters of georelief on spatial diferentation of landscape-ecological processes. Ecology (CSFR), Bratislava 10(2):115–158
Leser H (1991) Landschaftsökologie: Ansatz, Modelle, Methodik, Anwendung. Eugen Ulmer Verlag, Stuttgart, p 647
Leser H, Klink H J (1988) Handbuch und Kartierenleitung. Geoökologische Karte 1:25000 (KA GÖK 25). Forschungen zur Deutschen Landeskunde BD 228
Malík P, Bačová N, Hronček S, Ivanič B, Káčer Š, Kočický D, Maglay J, Marsina K, Ondrášik M, Šefčík P, Černák R, Švasta J, Lexa J (2007) Zostavovanie geologických máp v mierke 1:50000 pre potreby integrovaného manažmentu krajiny. Bratislava: ŠGÚDŠ, Manuskript—archív Geofondu, 552 pp
Malík P, Bahnová N, Ivanič B, Kočický D, Maretta M, Špilárová I, Švasta J, Zvara I (2011) Komplexná geologická informačná báza pre potreby ochrany prírody a manažmentu krajiny. Záverečná správa + prílohy, 149 pp + prílohy, Bratislava: MŽP SR, ŠGÚDŠ
Mander Ü, Müller F, Wrbka T (2005) Functional and structural landscape indicators: upscaling and downscaling problems. Ecol Ind 5(4):267–272
Mazúr E, Lukniš M (1986) Geomorfologické členenie SSR a ČSSR. Časť Slovensko. Slovenská kartografia, Bratislava
Mazúr E et al (1980) Atlas Slovenskej socialistickej republiky. Slovenská akadémia vied, Slovenský úrad geodézie a kartografie. Bratislava: VEDA, 296 pp
Mazúr E, Drdoš J, Urbánek J (1983) Krajinné syntézy—ich východiská a smerovanie. Geografický časopis 35:3–19
Mičian Ľ, Zatkalík F (1990) Náuka o krajine a starostlivosť o životné prostredie. /Vysokoškolské skriptum/, KFG PRIF UK, Bratislava, 137 pp
Midriak R (1999) Ekologická únosnosť vysokohorských oblastí z hľadiska morfodynamických procesov. Životné prostredie, 33, 1, ÚKE SAV, Bratislava, 42–45
Miklós L (1991) Morphometric indices of the relief in the LANDEP methods and their interpretation. Ecology (ČSFR) 10:159–186
Miklós L, Izakovičová Z (1997) Krajina ako geosystém. VEDA, SAV, Bratislava, p 152
Miklós L, Špinerová A (2011) Krajinno-ekologické plánovanie LANDEP. VKÚ, Harmanec, p 159
Miklós L, Špinerová A (2013) Krajinné plánovanie. Technická univerzita Zvolen, Praktikum, p 96
Miklós L, Hrnčiarová T, Kozová M (1984) Priestorová štruktúra hydrologických systémov na vybranom území VSN. Sborník prací GÚ, 6, Brno: ČSAV, pp 54–59
Miklós L, Kozová M, Ružička M et al (1986a) Ekologický plán využívania Východoslovenskej nížiny v mierke 1:25,000 (Ecological management plan of the Eastern lowlands at a scale of 1:25,000, in Slovakian). In: Ekologická optimalizácia využívania VSN, 3:5–312. ÚEBE CVEB SAV. Bratislava

Miklós L, Miklisová D, Reháková Z (1986b) Systematisation and automatisation of decision-making process in LANDEP method. Bratislava, Ecology (ČSSR) 5(2):203–232

Miklós L, Matečný I, Kozová M (1991) Interpretation of the morphometric relief indices for spatial differentiation of microclimatic conditions. Ekológia (ČSFR), Bratislava 10(2):223–226

Miklós L, Krcho J, Hrnčiarová T, Matečný I, Kozová M (1990) Interpretácia morfometrických vlastností reliéfu v krajinnoekologickom plánovaní LANDEP. ÚKE SAV, Bratislava, p 94

Miklós L, Hrnčiarová T et al (eds) (2002) Atlas krajiny Slovenskej republiky. 1. vydanie, MŽP SR Bratislava, SAŽP Banská Bystrica, 344 pp

Miklós L, Izakovičová Z, Boltižiar M, Diviaková A, Grotkovská L, Hrnčiarová T, Imrichová Z, Izakovičová Z, Kočická E, Kočický D, Kenderessy P, Miklós L, Mojses M, Moyzeová M, Petrovič F, Špinerová A, Špulerová J, Štefunková D, Valkovcová Z, Zvara I, Hreško J, Lauko V, David S (2005) Atlas of representative geoecosystems of Slovakia. ÚKE SAV, Bratislava, p 123

Miklós L, Izakovičová Z, Boltižiar M, Diviaková A, Grotkovská L, Hrnčiarová T, Imrichová Z, Izakovičová Z, Kočická E, Kočický D, Kenderessy P, Mojses M, Moyzeová M, Petrovič F, Špinerová A, Špulerová J, Štefunková D, Valkovcová Z, Zvara I, Hreško J, Lauko V, David S (2006) Atlas of representative geoecosystems of Slovakia. ÚKE SAV, II. Doplnené vydanie, Bratislava, p 124

Miklós L, Diviaková A, Izakovičová Z (2011a) Ekologické siete a územný systém ekologickej stability. Vydavateľstvo TU vo Zvolene, 141 pp

Miklós L, Ivanič B, Kočický D (2011b) Krajinnoekologická základňa integrovaného manažmentu povodia Ipľa. Digitálna databáza a tematické mapové vrstvy. Projekt HUSK 0801/2.1.2/0162 Vytvorenie jednotného monitoringu na báze priestorového informačného systému v povodí Ipľa. Banská Štiavnica, Esprit, spol. s r. o. (elektronický zdroj)

Miklós L, Németh R, Verrasztó Z (2014) Application of GIS in studying the drainage basin of the Ipoly River. (Scientific Annals of the Danube Delta Institute Tulcea, Romania. 20:109–128

Minár J (1992) The principles of the elementary geomorphological regionalization. Acta Facultatis Rerum Naturalium Universitas Comenianae, Geographica, Bratislava: PRIF UK 33:185–198

Minár J (1994) Niektoré teoreticko-metodologické problémy geomorfológie vo väzbe na tvorbu komplexných geomorfologických máp. Kandidátska dizertačná práca, Bratislava: PRIF UK, 214 pp (nepubl.)

Minár J (1995) Geografický prístup k výskumu teritoriálnej diferenciácie litosféry. In: Trizna M (ed) Vybrané problémy súčasnej geografie a príbuzných disciplín. KFGaG PRIF UK, Bratislava, pp 75–81

Minár J (1998) Georeliéf a geoekologické mapovanie vo veľkých mierkach./Habilitačná práca/, Bratislava: PRIF UK, 166 pp (nepubl.)

Minár J, Tremboš P (1994) Analýza georeliéfu ako podklad pre komplexný krajinno-ekologický výskum (modelové územie "Rudno"). Acta Facultatis Rerum Naturalium Universitas Comenianae, Geographica, PRIF UK, Bratislava 35:35–49

Minár J et al (2001) Geoekologický (komplexný fyzickogeografický) výskum a mapovanie vo veľkých mierkach. Geografické štúdie, PRIF UK, Bratislava, p 209

Mosimann T (1984) Methodische Grundprinzipien für die Untersuchung von Geoökosystemen in der topologischen Dimension. In: Geomethodica, Veröff. 9. Basler Geometh Coll, Bd. 9:31–65

Mosimann T (1990) Ökotope als elementare Prozesseinheiten der Landschaft (Konzept zur prozessorientierten Klassifikation von Geosystem). Geosynthesis, 1, Geographisches Institut der Universität Hanover, 56 pp

Ružička M, Miklós L (1982) Landscape ecological planning (LANDEP) in the process of territorial planning. Ekologia (CSSR) 1(3):297–312

Ružička M, Miklós L (1990) Basic premises and methods in landscape ecological planning and optimization. In: Zonneveld IS, Forman RTT (eds) Changing landscapes: an ecological perspective. Springer–Verlag, New York, pp 233–260

Ružička M, Ružičková H (1978) Žigrai F (1978) Krajinné zložky, prvky a štruktúra v biologickom plánovaní krajiny. Quastiones Geobiologicae 23:7–63

Ružičková H, Ružička M (1973) Druhotná štruktúra krajiny ako kritérium biologickej rovnováhy. Quaestiones Geobiologicae 12:23–62

Sochava VB (1978) Vvedenie v učenije o geosistemach (An Introduction to the Science of Geosystems in Russian). Nauka, Novosibirsk, p 319

Špinerová A (2010) Krajinno-ekologické limity poľnohospodárskeho využitia povodia Ilijského potoka. VKÚ, a. s., Harmanec, 118 pp

Szabó Sz, Csorba P, Varga K (2008) Landscape indices and landuse—tools for landscape managment. In: Methods of Landscape Research, Plit J, Andreychouk V (ed) Dissertations commission of cultural landscape, vol 8, pp 7–20, ISSN 1896-1460, Sosnowiec, Poland

Tremboš P (1991) Kritériá a vymedzenie základných krajinných jednotiek. KFGaG, Prif UK, Bratislava, 15 pp (nepubl.)

Tremboš P (1994) Geoekologická syntéza Diviackej pahorkatiny—metodický postup vymedzovania základných krajinných jednotiek. In: Minár J (ed) Prírodná časť krajiny, jej výskum a návrhy na využitie, /Zborník/. KFGaG PRIF UK, Bratislava, pp 118–120

Verrasztó Z (1979) Land formation and the geological aspects of environmental protection. In: Symposium Changes of the geological environment under the influence of man´s activity. IAEG National group, Krakow-Sandomierz-Belchatow-Plock-Warszawa, 135–141

Verrasztó Z (2010) Környezeti monitoring vizsgálatok az Ipoly vízgyűjtőjén (Tájökológiai Lapok. Gödöllő 8(3):532–561

Chapter 4
Characteristics of the Indicators of Geocomplexes

Abstract The chapter presents the set of detailed indicators of the geocomplexes including the tables of domain values of these indicators often used in landscape-ecological studies. Chapter includes the description, definition and characteristics of indicators of all three structures of the landscape as:

– indicators of properties of the primary landscape structure as: indicators of sub-complex georelief, geological base (bedrocks) and soil-forming substrate, waters, soils, climate, potential nature vegetation,
– indicators of properties of current landscape structure as: indicators of sub-complex current landscape structure and land cover, biotic-anthropic complex, real vegetation and habitats,
– indicators of properties of tertiary landscape structure as: sub-complex socio-economic factors for nature and landscape conservation, for protection of natural resources, of urbanisation, industry, agriculture production, transport and communal activities, sub-complex bound to the deterioration of the environment, sub-complex of geodynamic phenomena with the character of stress factors and sub-complex of administrative and sectoral boundaries.

A substantial part of the chapter is the tables with the domain values of indicators. The chapter is illustrated by 25 tables.

Keywords Indicators · Sub-complexes · Domain values

As described above the synthetic, well defined and precisely mapped geocomplexes are considered the **basic operational units** of the decision-making process for integrated landscape management. For this reason, their properties/attributes, indicators of properties and values of indicators must be arranged in a mode suitable for multipurpose use. According to previous experiences the next chapters presents the set of detailed characteristics of the geocomplexes which are often used in landscape-ecological studies.

L. Miklós et al., *Landscape as a Geosystem*,
https://doi.org/10.1007/978-3-319-94024-3_4

The structured character of the data enables differentiation of analytical and synthetic indicators of the various landscape structures: indicators of the properties of the primary landscape structure—abiocomplexes and physical-geographic complexes, indicators of the secondary landscape structure—biotic-anthropogenic complexes, indicators of the tertiary landscape structure—the socio-economic complex.

4.1 Indicators of Properties of the Primary Landscape Structure

The primary landscape structure is characterized by selected indicators of the properties of **georelief, bedrock, soil, climate and groundwater**. As a complex interpreted characteristic of abiocomplexes and theoretical knowledge about the ecological requirements of plant communities this complex also includes characterization of **potential natural vegetation**.

The resulting model of the abiocomplex with selected features of sub-complexes can be written as follows:

$$\mathbf{ABC} = (\mathbf{Re1}, \mathbf{Re2}, \mathbf{Re3}, \mathbf{Re4}, \mathbf{Re5}, \mathbf{Re6}, \mathbf{Ge1}, \mathbf{Ge2}, \mathbf{Wa1}, \mathbf{So1}, \mathbf{So2}, \mathbf{So3}, \mathbf{So4}, \mathbf{Cl1}, \mathbf{Cl2})$$

Individual elements/sub-complexes of the abiocomplex (ABC) contain the characteristics of the following properties:

- **sub-complex Re-relief**:
  - morphographic-position characteristics of form of georelief—Re1;
  - topographic position—Re2;
  - slope angle—Re3;
  - orientation of the relief to the cardinal points (aspect)—Re4;
  - profile curvature—Re5;
  - horizontal curvature—Re6;
  - choric characteristics of the relief—contributing area (microcatchment)—Re7;
  - choric characteristics of the relief—inclusion in the detailed hydrological catchment—Re8;
- **sub-complex Ge—geological base (bedrocks)—soil-forming substratum complex**:
  - type of geological substrate—Ge1;
  - quaternary cover thickness—Ge2;
- **sub-complex Wa—groundwater**:
  - depth of level of groundwater—Wa1;

- **sub-complex So—soil**:
  - soil subtype—So1;
  - soil texture (grain size)—So2;
  - depth of soil—So3;
  - skeletality of soil—So4;
- **sub-complex Cl—relief-climate**:
  - insolation of the georelief—Cl1;
  - climate geographical region—Cl2.

A more detailed description of individual indicators of sub-complex properties is as follows:

4.1.1 Sub-complex Re: Georelief

- The primary spatial unit for creating a database of abiocomplexes is morphological-morphographic-position types of georelief, in the boundaries, onto which additional selected characteristics were added gradually, namely: morphological-morphographic-position type—Re1;
- topographic position (relating to transport of material)—Re2;
- slope angle—Re3;
- orientation of the relief to the cardinal points (aspect)—Re4;
- profile curvature—Re5;
- horizontal curvature—Re6.

R1, R2—Morphological-Morphographic-Position Type of Georelief

The georeference layer—**morphological-morphographic-position types of georelief**—have been created by computer-aided visual on-screen analysis of the contour lines. Based on the above properties morphotopes are delimited. Across Slovakia more than 500.000 polygons of morphotopes have been allocated, they were divided into the following categories (Table 4.1).

Re3—The Slope, Re4—Dominant Orientation of Georelief

A 20 × 20 digital elevation model was derived from the vector representations of height field and the "skeleton" of relief. The height field was represented by contour lines constructed on the basis of the point grid in spatial accuracy of basic map (BM) 1:10,000. The skeleton of relief was defined as an oriented river network (river networks respecting the rules of network topology) in positional accuracy corresponding to ZB GIS.

Morphometric characteristics (elevation, orientation of slope to cardinal points, profile and horizontal curvature, dissection, insolation) were generated from 20 × 20 DTM and each polygon of abiocomplexes is characterized by the average or

Table 4.1 Domain values of attribute of morphological-morphographic-position type

Code/no.	Morphological–morphographic description—Re1	Position—Re2
01	Flat top	Initial
02	Plateau top	Initial
03	Dome-shaped top	Initial
04	Ridge	Initial
05	Compact dell	Transport
06	Ground of dell valley	Transport
07	Ground of slope valley	Transport
08	Compact slope valley	Transport
09	Transport slope	Transport
10	Slope platform	Transport
11	Large accumulation plain	Transport
12	Wavy plain of sand dune	Transport
13	Significantly wavy slope	Transport
14	Slope divided by gullies and small valleys	Transport
15	Terrace (river)	Transport
16	Narrow floodplain of mountain streams	Transport/final
17	River channel in section (natural and anthropogenic)	Transport/final
18	More significant elevation as part of floodplain	Final
19	Slope in foothill position/foothill	Final
20	Alluvial fan	Final
21	Wide riverside floodplain	Final
22	Closed depression whole (a-closed, b-oxbow)	Final
23	Bottom of waterlogged depression	Final
	Specific forms of relief	
24	Peak	Initial/temporary
25	Mountain saddle	Temporary
26	Sliding slope	Transport/temporary
27	Terraced slope	Transport/temporary
28	Scree and deposit of stones (stone sea)	Transport/temporary
29	Rock scarp	Transport/temporary
30	Bottom of glacial trough	Transport/temporary
31	Gully to ravine	Transport/temporary
32	Moraine mound	Transport/temporary
33	Glacial cirque	Final/temporary
34	Bottom of water reservoir	Final/temporary
35	Anthropogenic form	Undefined/temporary

Table 4.2 Domain values of attribute of slope angle

Code/no.—Re3	Interval of slope angle (°)
1	0–1
2	1–3
3	3–7
4	7–12
5	12–17
6	17–25
7	25–35
8	35–90

Table 4.3 Domain values of attribute of orientation of the relief to the cardinal points (aspect)

Code/no.—Re4	Interval of azimuth (°)
1	337.5–22.5
2	22.5–67.5
3	67.5–112.5
4	112.5–157.5
5	157.5–202.5
6	202.5–247.5
7	247.5–292.5
8	292.5–337.5

Table 4.4 Domain values of the attribute of profile curvature in the direction of gradient curves

Code/no.—Re5	Normal form
1	Concave
2	Linear
3	Convex
4	Complex

prevailing value (in work also the minimum and maximum), values were reclassified into intervals.

The dominant **slope angle** was set in traditional interval values (Table 4.2).

Dominant orientation is expressed as a stable indicator, on the basis of which we can calculate various real and average values of exposure of the relief to the sun. It thereby becomes an important morphoclimatic parameter. Traditional interval values were considered (Table 4.3).

Re5, Re6—Curvature of Georelief

Profile (or normal) curvature characterizes the size of flexure of gradient curves. It represents the local effect of the spatial variation of the inclination of relief. It is a crucial indicator for determining the tendency of the movement of water and material down the slope (acceleration, deceleration). It represents the curvature of the gradient curves. These traditional shapes were considered (Table 4.4).

Table 4.5 Domain value of attribute of horizontal curvature in the direction of contour lines

Code/no.—Re6	Horizontal form
1	Concave
2	Linear
3	Convex
4	Complex

As normal curvature of gradient lines expresses the local slope change and the related flow characteristics, **horizontal curvature** characterizes the curvature of contour lines. It is a crucial indicator for determining flow direction and integration of water (concentration, scattering) on the slopes. It represents both the direction of movement of the material, which lead to the pooling—convergence (concave curvature with a negative value of curvature) or distraction—divergence (in a convex shape with positive values of curvature) of surface runoff (Table 4.5).

Re7, Re8—Choric Characteristics of Georelief

These parameters constitute separate data layers, they are choric characteristics of relief:

- **Re7**—contributing area (microcatchment);
- **Re8**—inclusion in the detailed hydrological catchment.

These indicators in simplified form characterize the **horizontal relationships** of geocomplexes and are useful for a variety of landscape-ecological evaluations.

Re7—contributing area (microcatchment) is the size of the area from which the water flows over the surface during rainfall events to a defined point of relief. It is a useful morphometric indicator, used for rapid estimation of the size (mass) of surface water flow at a given point on the surface. It is expressed in m^2. Microcatchments of more than 100 m^2 have been delimited throughout Slovakia.

The microcatchment has been derived from DTM in GIS and raster representation of the runoff. For the purposes of calculation, two layers were derived: representative of the multidirectional orientation of the range (0–360°), and an eight-direction outlet orientation. The microcatchment was calculated from an algorithm on the resulting vector—Dinf (Tarboton 1997), so that the derived river network respects the current course of the river network in ZB GIS. The run-of network is ensured by integrating rasterized layers representing the river network, so that lower orders of flow interfere with their values (pixel values) to higher order flows. The calculation was executed in the raster format GeoTIFF the programme Taudem allowing Multithreaded computation over large raster data sets. The calculation has been set up to the slopes of the convex surface of the diffusion flux is used and the algorithm DINFO river bed again D8 algorithm, which ensures convergence of the flow (preventing diffusion when it has to be concentrated) (Malík et al. 2011).

Re8—detailed hydrological catchments (microbasins) are the smallest spatial units defined under the Water Management Map of the Slovak Republic at 1:50,000. Detailed basins are hierarchically the lowest territorial hydrological unit within the higher classes, which are the **basic basins** (in Slovak there are 55 basic basin areas), above them there are partial-basins (Danube, Morava, Vah, Hron, Ipeľ, Slaná, Bodrog, Hornád, Bodva, Dunajec and Poprad). The highest units of hydrological regionalization are the **main basins**—the Danube and the Vistula. In Slovakia there are 4337 detailed hydrological catchments with associated hydrological numbers of catchment (HNC). Each detailed hydrological catchment has the name of the water flow and hydrological number (e.g. 3-01-02-068 Žákovský stream in the cadastral area of Veľká Lomnica, Huncovce).

Detailed catchments were delimited by computer-assisted visual on-screen interpretation of maps and then topologically adjusted for elevation, river network and other relevant objects ZB GIS (dams, the road network, water bodies, etc.). The GIS Spatial Analyst tool was used.

4.1.2 Sub-complex Ge: Geological Base (Bedrocks)—Soil-Forming Substrate Complex

The geological base (bedrocks)—soil-forming substrate complex is uppermost relatively thin part of the geological substrate, usually of quaternary age, which is the most important part of the solid ground for practical landscape-ecological research (Miklos and Izakovičová 1997, modified).

The database of abiocomplexes, given by the nature and objective of the research projects for which has been compiled, contains primary and derived geological data, namely

(a) lithological-genetic characteristics of the base of abiocomplexes:

- genetic characteristics of the base of abiocomplexes;
- lithological composition of the base of abiocomplexes;

(b) characteristics of quaternary geological substrate complex:

- genetic type of quaternary sediments;
- thickness of quaternary cover;

(c) Engineering-geological properties of abiocomplexes:

- engineering-geological region and sub-region;
- predominant granularity composition of the substratum (priority upper horizons)—classification of rock and semi-rock by the strength level, prevailing plasticity of fine-grained soils, prevailing consistency of fine-grained soils and prevailing settling of gravel and sand in the context of the valid Slovak Technical Norm (STN) 72 1001;
- behaviour of rocks and soil in contact with water (type and the intensity of transmission, susceptibility to frost, capillary action, volume changes (retraction, swellability, breaking) applicable by STN 72 1002);

- susceptibility to settling of loess sediments applicable by STN 73 1001.

From the scope of data for landscape-ecological purposes we suggest to consider primarily the following indicators:

- type of geological base (bedrock)—soil-forming substrate complex—Ge1;
- thickness of quaternary sediments—Ge2.

The selected indicators of the geological substrate complex were prepared taking into account the needs of method of Landscape Ecological Planning LANDEP. Other indicators can be used in specialized studies and projects.

Ge1—Type of Geological base (Bedrock)—Soil-Forming Substrate Complex

This indicator has been fully processed, even in areas where the quaternary has not been mapped. Values have been attributed to individual polygons of abiocomplexes based on the characteristics of the relief, the genetic type of quaternary sediments and the geological substrate. Subsequently, this classification was used in typification of abiocomplexes. The following categories were considered (Table 4.6).

Ge2—Thickness of Quaternary Sediments

As part of the digital database of abiocomplexes the average thickness (depth) of quaternary sediments is reported. The thickness of the quaternary sediments was derived on the basis of isolines of thickness of the quaternary deposits. For compilation comprehensive geological research was carried out over the whole territory of Slovakia, which was supplemented by data from hydrological data (HG) and of engineering wells, wells and other technical work. In addition to this database the data was also used from geophysical and quaternary geological sections, as well as other available data (texts to basic and applied geological maps of individual regions, a review of previously compiled maps of thicknesses of quaternary on larger scales, other professional papers and reports dealing with the issue of Quaternary DTM, aerial photographs). Isolines of thickness are digitally processed at defined intervals of 0–2 m; 2.1–5 m; 5.1–10 m; 10.1–15 m; 15.1–20 m; thickness ranging from 21 to 100 m at intervals of 10 metres and a thickness above 100 m at intervals of 50 m. The layer respects the current DTM, as well as boundary areas of occurrence of relevant genetic types of sediments (Table 4.7).

4.1.3 Sub-complex Wa: Ground Water

The Slovak database of abiocomplexes contains the following hydrogeological data:

- hydrogeological index of the character of the rock;
- filtration coefficient
- transmissivity coefficient (flow capacity);
- saturation coefficient of free water level;

Table 4.6 Characteristics of domain of geological base (bedrock) – soil-forming substrate complex

Code/no.—-Ge1	Genetic group of sediments	Type of geological substrate complex
01	Fluvial sediments	Fluvial sediments with organic addition
02	Fluvial sediments	Clay-loamy to clay floodplain sediments
03	Fluvial sediments	Loamy to sandy-loamy floodplain sediments
04	Fluvial sediments	Sandy to loamy-sandy floodplain sediments
05	Fluvial sediments	Sandy-loamy terrace sediments
06	Fluvial sediments	Gravel-loamy to gravel floodplain sediments
07	Fluvial sediments	Gravel-loamy terrace sediments
08	Proluvial sediments	Sandy-loamy to sandy proluvial sediments
09	Proluvial sediments	Gravel-loamy to gravel proluvial sediments
10	Aeolic sediments	Loess to loess-loam
11	Aeolic sediments	Drifting sands
12	Deluvial sediments	Clay to sandy-loamy deluvial sediments
13	Deluvial sediments	Gravel-loamy deluvial sediments
14	Glacifluvial sediments	Sandy-gravel glacifluvial sediments
15	Glacifluvial sediments	Gravel-boulder glacifluvial to glacial sediments
16	Glacifluvial sediments	Boulder glacial sediments (moraine)
17	Other sediments	Gravel-boulder colluvial sediments
18	Other sediments	Organogenic sediments (peat)
19	Other sediments	Travertine
20	Other sediments	Mixed landslide materials
21	Other sediments	Anthropogenic sediments
22	Local weatherings and deluviums (eluvium, deluvium) of rocks or protruding bare rocks	Silt-loam to clay weathering in non-hardened clay sediments (tertiary clay and silt clay)
23	Local weatherings and deluviums (eluvium, deluvium) of rocks or protruding bare rocks	Gravel-sand in non-hardened gravel sediments (tertiary gravels with deposits of sands)
24	Local weatherings and deluviums (eluvium, deluvium) of rocks or protruding bare rocks	Loamy weatherings in hardened clay-silt rocks (shales, clay, limestone and marlstone)
25	Local weatherings and deluviums (eluvium, deluvium) of rocks or protruding bare rocks	Loamy weatherings in flyschoid rocks (shales, sandstone, limestone and marlstone)

(continued)

Table 4.6 (continued)

Code/no.—Ge1	Genetic group of sediments	Type of geological substrate complex
26	Local weatherings and deluviums (eluvium, deluvium) of rocks or protruding bare rocks	Loamy weatherings to loam-stone debris on sandstone-conglomerate rocks (sandstone, conglomerate, shales, quartzite, arkoses, limestone and marlstone)
27	Local weatherings and deluviums (eluvium, deluvium) of rocks or protruding bare rocks	Loamy to clay-rock weatherings on massive limestone-dolomite rocks (limestone, dolomite)
28	Local weatherings and deluviums (eluvium, deluvium) of rocks or protruding bare rocks	Loamy to loamy-stone weatherings on volcanic pyroclastics (tuff, tuffs)
29	Local weatherings and deluviums (eluvium, deluvium) of rocks or protruding bare rocks	Loamy-gravel to stone weatherings on effusive rocks (melaphyre, rhyolite, rhyodacites, andesite, basalite)
30	Local weatherings and deluviums (eluvium, deluvium) of rocks or protruding bare rocks	Loamy weatherings to loam-stone debris on intrusive magmatic rocks (granites, granodiorite, diorite)
31	Local weatherings and deluviums (eluvium, deluvium) of rocks or protruding bare rocks	Loamy, gravel to stone weatherings on low metamorphic rocks (phyllite, phyllonite)
32	Local weatherings and deluviums (eluvium, deluvium) of rocks or protruding bare rocks	Loamy, gravel to stone weatherings on highly metamorphised rocks (gneiss, mica schist, amphiboles, migmatites)

- average depth of the ground water level—Wa.

The selection of most of the above parameters is based on the "design methodology for geographic information system in the project POVAPSYS", implemented by the Slovak Hydrometeorological Institute in 2005–2006. The digital map was based on the processing of a large database of hydrogeological wells in the archive of the Geofond. At the time of evaluation, the Geofond, for the whole territory of Slovakia, had 22,778 record sheets of hydrogeological boring.

For the needs of landscape-ecological evaluation and management, the depth of the groundwater level is of the utmost importance. We therefore did not consider other indicators in compiling the targeted geocomplexes.

*Wa*1—*Depth of Level of Underground Water Under Terrain*

Completion of the data on the groundwater table for the whole territory of Slovakia was one of the most important benefits of the project of Malik et al. (2011). The researchers of the project however strongly warn potential users of data on height levels of ground water below the surface of the relief that it is totalised data compiled in order to obtain a nationwide overview. In any case, it is not possible to use this

Table 4.7 Thickness of quaternary sediments

Code/no.—Ge2	Interval of depth of quaternary (m)
01	1–2
02	2–5
03	5–10
04	10–15
05	15–20
06	20–30
07	30–40
08	40–50
09	50–60
10	60–70
11	70–80
12	80–90
13	90–100
14	100–150
15	150–200
16	200–250
17	250–300
18	300–350
19	350–400
20	400–450
21	450–500

data for studies on detailed regional or local level—in this case it is necessary to proceed with the incorporation of available additional data from other (geological-engineering) sensors and in particular to take into account the time of variation of groundwater levels associated with the climate cycle in the country.

Data on the levels of ground water below the surface were obtained from archived hydrogeological boreholes (with a record of the depth of the groundwater level, there were at the time of solution, a total of 16,784) and from processed geo-engineering probes (total 17,169) (Geofond SGIDŠ). Of these hydrogeological boreholes, wells were excluded with forced (artesian) groundwater level, so processing took place on 16,380 point data items from hydrogeological wells. In Geofond SGIDŠ there are archived tens of thousands of reports of an engineering nature that describe hundreds of thousands—an estimated 7000 engineering boreholes (Malík et al. 2011). Probes were selected containing an indication of the depth of the groundwater level that could be located in the S-JTSK system and the Balt height system (B.p.v.) on a scale of 1:50,000. It was necessary to create a single data structure, consisting of no more than a simplified framework to give the coordinates of the probe, its altitude, its designation, and depth and data.

Data on engineering-geological probes was saved using an MS Excel spreadsheet. Data was taken from archived reports, a total of 4734, representing 29.55% of all data on the depth of the groundwater level. For other territories, untouched required data and in territories of built of the rock environment of the type of a hydrogeological massif correlation of data was performed with morphometric characteristics of relief in abiocomplexes. On the basis of the forecast of the average ground water level, for each polygon of the abiocomplex, the hydromorphometric value of the index of concentration of data was extrapolated to the level of a grid of 200 × 200 m. For Karst rocks, or abiocomplexes formed from a geological point of view in the Middle and Upper Triassic limestone and dolomite their relationship with the local base level was investigated, the envisaged passage of the groundwater level was explicitly set at 2 m above the local erosion base. And groundwater levels in Karst rock environment established in that way were then extrapolated to a grid with dimensions 200 × 200 m.

Abiocomplexes were assigned a corresponding circulation type groundwater. Thus, they were allocated to individual areas for which the depth was determined using a methodology best capturing the patterns of circulation of ground water in the given circulation type of aquifer and at the same time taking into consideration the density of wells in different areas. This resulted in a final set of validated wells with a total of 27,688 points. For these points, a digital elevation model was used to calculate the attribute of altitude of the well. By subtracting the depth below the surface level from the altitude of borehole, for each well the elevation of groundwater level was determined. It was an analysis of the number of wells in regions with different types of circulation and has generated raster groundwater levels for each region by circulating types of aquifer (Bahnová et al. In Malik et al. 2011). The generation of the raster of ground water level for each region by circulation types of aquifers took place in the following five ways:

- by interpolating the ground water level of the alluvial areas and selected areas close to the alluvium;
- by interpolating groundwater levels for floodplains of mountain streams;
- by interpolating ground water levels in other areas with sufficient density of hydrological boreholes;
- by interpolating groundwater level for areas with a hydrogeological massif type of circulation for crystalline rocks, neovolcanites and flysch sediments;
- by interpolating ground water levels of areas with Karst and fissure-fissure groundwater circulation.

Subsequently the general directions of groundwater flow were identified. The procedure consisted of assessing the essential features of hydrogeological conditions, in determining the extent of effective precipitation (precipitation totals the difference and real evapotranspiration of the) extent of replenishment of ground water, identifying the areas of drainage of ground water and analysis of the identified direction of flow in existing studies. For the calculation of real evapotranspiration, the method of residual (detrended) Kriging was applied, linear stepwise regression (e.g. Thornton et al. 1997; Sitková 2000; Tveito 2002). The calculation of potential evapotranspiration was used as per Thornthwaite (1984). The raster map of the underground drainage was compiled from raster effective precipitation, by comparing the hydrogeological conditions of the territory to the map of groundwater flow (Krásný et al. 1981). This data was correlated and calibrated (Malík and Švasta In Malík et al. 2011).

Maps of groundwater flow direction were generated by extrapolating the information contained in hydrogeological maps of all levels (but especially 1:20,0000 and 1:50,000)—basic hydrogeological maps, special purpose hydrogeological maps generated within the research and detailed hydrogeological surveys archived in the Geofond SGIDŠ as well as purposed hydrogeological maps that are part of the complex maps of geofactors of the environment. In their processing it was necessary to conduct rectification of maps, their georeferencing, vectorization of hydroisohypses and vectorization of the maps referring to groundwater flow direction. The relevant structure of the data layer as attributes in addition to high levels (for hydroisohypses) and azimuth (for flow directions) reported the references to the relevant final report, its author, the year of submission and archival number under which the message is stored in Geofond SGIDŠ. The final form of flow direction was formed after supplementing and comparing data on runoff of groundwater (underground runoff) compiled from the raster of effective precipitation and comparison with the map of groundwater flow (Krásný et al. 1981). The map of directions of groundwater flow for the entire Slovak territory from the underlying data was also designed as a vector field for a raster of 200 × 200 m using a digital terrain model (DTM).

The indicator **Depth of level of ground water** under the surface took values from 0.1 to 675.6 m below ground level. For defining geosystems of the Slovak Republic they were incorporated into the following intervals (Table 4.8):

Table 4.8 Category of depth of level of ground water under the surface

Code/no.—Wa1	Depth of level of ground water (m)
1	<1
2	1–2
3	2–5
4	5–10
5	10–25
6	25–50
7	50–100
8	100>

4.1.4 Sub-complex So: Soil

In this place we will characterize only indicators of the basic properties of soil, which are:

- soil type and subtype;
- soil texture (grain size);
- depth of soil;
- skeletality of soil.

However, the database of abiocomplexes also contains information on other indicators on soil characteristics, namely

- coefficient of soaking (rate of the process of water into dry soil);
- permeability—hydraulic conductivity (the speed of movement of gravity water in saturated soil);
- full water capacity (maximum amount of water that soil can attract);
- retention of soil (potential amount of water, which the abiocomplex can hold);
- total flow capacity (parameters expressing integrated flow capacity).

For the purposes of integrated landscape management, the indicators most often used are the following:

So1 —Soil Types and Subtypes

Characteristics of **soil types and subtypes** were transposed from a digital map of soil types at 1:50,000 (Malík et al. 2007). The soils were classified in terms of morphogenetic classification system of soils of Czechoslovakia (Hraško et al. 1991; Collective 2000). The soils were not determined for the areas where the surface is formed from rock or water (Table 4.9).

So2—Soil Texture (Grain Size)

Based on available data, a digital map of soils grain size was compiled. The layer of soil grain size originated from the input data contained in the soil probes (grain-size fractions—sand, silt, clay in terms of the textural triangle (Hraško et al. 1991; Collective.... 2000) on a comprehensive survey of soil by complex soil survey (KPP) provided by the Soil Science and Conservation Research Institute (SSCRI) in the years 1961–1970 (Skalský and Balkovič 2002), for particular regions from the maps of Geofactors (Bodiš et al. 1998), for particular regions from the maps of Geofactors (Bodiš et al. 1998), from the Geochemical atlas of Slovakia (Rapant et al. 1996—about 18,000 grain-size analysis), from maps of forest soils from National Forest Institute (NFC). About forest soil there is less data, about 3500 grain-size analyses.

To derive maps of soil grain-size multiple linear regression and interpolation in geographically homogeneous medium were used with sufficient density of points of entry from the field. Geostatistical analysis was done on the entered values of those properties of geocomplex that would affect soil grain size—mainly the geological

Table 4.9 Categories of soil types and subtypes

Code/no.—P1	Soil subtype description
01	Lithic Leptosol (typical)
02	Haplic Regosol (typical)
03	Haplic Regosol (psephite)
04	Haplic Regosol (arenic)
05	Haplic Regosol (pelic)
06	Haplic Leptosol (typical)
07	Cambic Leptosol
08	Andic Leptosol
09	Leptic Podzol
10	Rendzic Leptosol (typical)
11	Rendzic Lithic Leptosol
12	Cambic Rendzic Leptosol
13	Histic Rendzic Leptosol
14	Rendzic Leptosol (skeletic)
15	Rendzic Leptosol (rubeficated)
16	(Para)rendzic Leptosol (typical)
17	Leptic Cambisol (calcaric)
18	Stagnic Rendzic Leptosol
19	Mollic Vertisol (typical)
20	Haplic Chernozem (typical)
21	Haplic Chernozem (arenic)
22	Haplic Chernozem (pelic)
23	Luvic Chernozem
24	Mollic Chernozem
25	Stagnic Chernozem
26	Mollic Fluvisol (typical)
27	Mollic Fluvisol (arenic)
28	Mollic Fluvisol (pelic)
29	Mollic Chernozem
30	Gleyic Mollic Fluvisol
31	Histic Gleyic Mollic Fluvisol
32	Cutanic Luvisol (typical)
33	Cutanic Luvisol (arenic)
34	Albic Cutanic Luvisol
35	Gleyic Cutanic Luvisol
36	Cutanic Luvisol (rubeficated)
37	Cutanic Albic Luvisol (typical)
38	Cutanic Albic Luvisol (arenic)

(continued)

Table 4.9 (continued)

Code/no.—P1	Soil subtype description
39	Fragic Cutanic Albeluvisol
40	Haplic Cambisol (typical)
41	Haplic Cambisol (psephitic)
42	Haplic Cambisol (arenic)
43	Haplic Cambisol (pelic)
44	Cambisol (Calcaric)
45	Cambisol (Eutric)
46	Cambisol (Dystric)
47	Andic Cambisol
48	Leptic Luvisol
49	Stagnic Cambisol
50	Haplic Cambisol (rubeficated)
51	Fulvic Andosol (typical)
52	Albic Podzol (typical)
53	Entic Podzol
54	Histic Podzol
55	Haplic Planosol (typical)
56	Luvic Planosol
57	Haplic Stagnosol
58	Gleyic Albeluvisol
59	Haplic Gleysol (typical)
60	Haplic Gleysol (arenic)
61	Histic Gleysol
62	Histosol (typical, dystric)
63	Histic Gleysol
64	Haplic Fluvisol (typical)
65	Haplic Fluvisol (psephitic)
66	Haplic Fluvisol (arenic)
67	Haplic Fluvisol (pelic)
68	Gleyic Fluvisol
69	Haplic Solonetz
70	Anthrosol (according to soil subtype, typical)
71	Anthrosol (according to soil subtype, degraded)
72	Technosol (typical)
73	Technosol (degraded)

Table 4.10 Categories of soil texture (grain size)

Code/no.—So2	Soil group	Granularity group
01	Sand	Light soil
02	Loam–sand	Light soil
03	Sand–loam	Moderately-heavy soil
04	Loam	Moderately-heavy soil
05	Silt–loam	Moderately-heavy soil
06	Silt	Moderately-heavy soil
07	Sand–clay–loam	Moderately-heavy soil
08	Clay–loam	Moderately-heavy soil
09	Silt–clay–loam	Moderately-heavy soil
10	Sand–clay	Heavy soil
11	Silt–clay	Heavy soil
12	Clay	Heavy soil

Table 4.11 Categories of soil depth

Code/no.—So3	Soil depth (cm)	Description of category of soil
1	Over 61	Deep
2	31–60	Moderately deep
3	Up to 30	Shallow

substrate complex, the slope angle of the relief, absolute and relative altitude, contributing area. They determine the respective shares of granularity fractions under which the triangle of soil granularity determines the categories of soil texture.

All soil characteristics were converted into abiocomplexes and then adjusted to take account of determined spatial links between soil grain size and other elements of the geocomplex, so as to ensure logical consistency of data. Traditional grain-size categories were considered (Table 4.10).

So3—Depth of Soil

For the needs using indicators of the properties of soils especially critical is depth and skeletality of soil. **Soil depth**—the depth of the soil profile, it expressed the traditional categories of agricultural land (this categorization has been applied also in forest soils) (Table 4.11).

So4—Skeletality (Graveliness and Stoniness) of Soil

Expressed as % skeleton share (particles with a diameter of more than 2 mm). These properties are derived from the soil types (in names that reflect initiality—leptosols) (Table 4.12).

Information on the depth and skeletality of the soils was taken from the maps of the comprehensive soil research realized by SSCRI in 1961–1970 and from Skalský and

Table 4.12 Categories of soil skeletality

Code/no.—So4	Skeletality of soil (%)	Description of category of soil
1	Up to 25	Without or scarcely skeletal
2	25–50	Moderately skeletal
3	Over 50	Very skeletal

Balkovič 2002), the forest maps (NFC), the work of Linkeš et al. (1988, 1997); Kobza et al. (2002). Properties were established as derived characteristics using knowledge on the links between spatial differentiation of soil cover and other landscape elements.

4.1.5 Sub-complex Cl: Relief—Climate

For landscape-ecological studies, the most important is the following indicator:

Cl1—Amount of the Sun Radiation on Georelief

In landscape-ecological studies, due to the difficulty of the experimental evaluation of microclimate conditions over large areas, as an indicator of temperature regime of the geosystem, **amount of the sun radiation on georelief** is used. It is the most important indicator of microclimate influenced by the spatial variability of the morphometrical conditions—mainly the slope angle and the orientation to the cardinal points. In studies of landscape it is used as an indicator of the temperature regime of the geosystem. Here it represents the average rate of global radiation (direct radiation + diffuse radiation) expressed in Wh m^{-2} (Table 4.13).

To create the sun radiation model, the GIS Arc Map Spatial Analyst extension was used, specifically the module to calculate the length and intensity of solar radiation. The impact of solar radiation on the territory was evaluated for various defined time intervals, taking into account atmospheric influences, latitude and altitude, slope, orientation, daily and seasonal changes in the declination of the sun, as well as the impact of shading the surrounding terrain. The input to the model is the digital terrain model DTM with resolution of 20 × 20 m derived from contour maps of BM 1:10,000, the channels/lakes network. The output of the model is the flow of global radiation (direct radiation + diffuse radiation) for each month and the whole year (Malík et al. 2011).

Cl2—Climate Geographical Type

The climate geographical types are formed by orographic units in three groups—group of lowland, basin and mountain climate. These are characterized by a set of climate indicators. Each type of ABC is included in a climate geographical type. The origin of the map of climate types (Lapin et al. In Miklós and Hrnčiarová (eds) et al. 2002) does not express in sufficient detail the heterogeneity of climatic elements mainly due to orographic effects. Therefore, for the purposes of this work it has been decided to use the climate geographical types from the Atlas of the Slovak Socialist Republic (Tarábek In Mazúr et al. 1980). Boundaries of sites cited on the above maps of climate geographical types were significantly modified and clarified on the basis of the digital terrain model a BM 1:10,000. In the next step, such specified characteristics of climate geographical sites of abiocomplexes were assigned by superposing onto each and were again evaluated based on altitude and morphological-position type. Synthesis was used to define morphoclimatopes that are homogeneous in terms of climatic conditions (Tables 4.14 and 4.15).

Table 4.13 Categories of the amount of the sun radiation on georelief

Code/no.—Cl1	Value of amount of sun radiation per year (kJ)
01	900 and less
02	900–950
03	950–1000
04	1000–1050
05	1050–1100
06	1100–1150
07	1150–1200
08	1200–1250
09	1250–1300
10	1300–1350
11	1350–1400
12	1400–1450
13	1450 and more

Table 4.14 Climate geographical types and subtypes

Code/no.—Cl2	Climatic-geographical type	Climatic-geographical subtype
01	Lowland climate	Warm
02	Lowland climate	Predominantly warm
03	Basin climate	Warm
04	Basin climate	Moderately warm
05	Basin climate	Moderately cold
06	Basin climate	Cool
07	Mountain climate	Warm
08	Mountain climate	Moderately warm
09	Mountain climate	Moderately cool
10	Mountain climate	Cool
11	Mountain climate	Cold
12	Mountain climate	Very cold

Table 4.15 Characteristics of domain values of climate geographical types

Code of climate geographical type	Climate geographical type	Climate geographical subtype	Total temp 10° and more	Lower interval of average January temps	Upper interval of average January temps	Lower interval of average July temps	Upper interval of average July temps	Lower interval of amplitude of average monthly temps	Lower interval of amplitude of average monthly temps	Lower interval of annual total rainfall	Upper interval of annual total rainfall
Climate geotype	Type	Subtype	t_sum	t_jan_min	t_jan_max	t_jul_min	t_jul_max	amp_min	amp_max	prec_min	prec_max
01	Lowland climate	Warm	3000–3200	−4	−1	19.5	20.5	22	24	530	650
02	Lowland climate	Mostly warm	2600–3000	−4	−1.5	18.5	19.5	21.5	24	650	700
03	Basin climate	Warm	2600–3000	−4	−2	20	18.5	22	24	600	700
04	Basin climate	Moderately warm	2400–2600	−5	−2.5	18.5	17	20	24	600	800
05	Basin climate	Moderately cool	2100–2400	−6	−3.5	17	16	20	24	600	850
06	Basin climate	Cool	1500–2100	−6	−4.5	16	14.5	20	22.5	610	900
07	Mountain climate	warm	2400–2900	−5	−2	19.5	17.5	21	23	600	800
08	Mountain climate	Moderately warm	2200–2400	−6	−3.5	17.5	17	21	23	650	850
09	Mountain climate	Moderately cool	1600–2200	−6	−4	17	16	21	21.5	800	900

(continued)

Table 4.15 (continued)

Code of climate geographical type	Climate geographical type	Climate geographical subtype	Total temp 10° and more	Lower interval of average January temps	Upper interval of average January temps	Lower interval of average July temps	Upper interval of average July temps	Lower interval of amplitude of average monthly temps	Lower interval of amplitude of average monthly temps	Lower interval of annual total rainfall	Upper interval of annual total rainfall
Climate geotype	Type	Subtype	t_sum	t_jan_min	t_jan_max	t_jul_min	t_jul_max	amp_min	amp_max	prec_min	prec_max
10	Mountain climate	Cool	1200–1600	−6.5	−5	16	13.5	19.5	21	800	1100
11	Mountain climate	Cold	500–1200	−7	−6	13.5	11.5	18	20	1000	1400
12	Mountain climate	Very cold	0–500	−11	−7	11.5	4	15.5	19	1200	2130

4.1.6 Sub-complex PNV: Potential Natural Vegetation

This is a separate data layer outside the database of abiocomplexes, for our needs designated as **PNV**.

Potential natural vegetation is such vegetation that would, given the climate, soil and hydrological conditions develop in a particular site, if the impact of human activity stopped immediately. It is interpreted indicator of the real abiocomplexes—a thought-construction for possible vegetation on sites of ABC, according to the theoretical knowledge on the ecological requirements of plant communities.

The category of communities of potential natural vegetation is based on the work Michalko et al. (1986) and has been processed and prepared in a vector format based on working manuscripts at 1:50,000, supplied by the department of geobotany at the Botanical Institute, Slovak Academy of Sciences (SAS), Bratislava (Table 4.16).

4.2 Indicators of Properties of Current Landscape Structure

We characterize the elements of the current landscape structure for practical purposes, mainly in terms of two aspects, as:

- **elements of land cover**: as a result of land use that we consider both as physiognomic-ecological formations, and at the same time as **mapping units of the current landscape structure**—CLS elements;
- **real vegetation**: with basic resolution, the mapping units of vegetation are identical to the elements of land cover, as well as the physiognomic-ecological formations with emphasis on the biotic content, therefore vegetation covered landscape elements are usually further divided. From this aspect, these factors may also be considered as **habitats** of flora and fauna, at the basic level as habitats of forest vegetation (forest habitats) and non-forest vegetation habitats (non-forest habitats).

From the geosystem perspective as well as from the perspective of creation of GIS, georeferencing elements of the current landscape structure are still elements of land cover, divided in varying degrees of detail. Their characteristics as CLS mapping units, element of land cover, land use element, physiognomic-ecological formation of real vegetation, habitat, or other indicators can be considered as indicators of individual characteristics of the elements of the current landscape structure.

Table 4.16 Categories of communities of potential natural vegetation

Code/no.—PNV	Potential natural communities
01	Willow-poplar floodplain woods of major rivers (soft wetlands)
02	Elm floodplain woods in the basins of major rivers (hard wetlands)
03	Submountain and montane alder woods
04	Alder fen woods
05	Hygrophilous lowland oak-hornbeam forests
06	Peri-pannonian oak-hornbeam forests
07	Carpathian oak-hornbeam forests
08	Mixed broadleaved-coniferous forest in the Northern Carpathian basin
09	Oak woods and oak woods with *Quercus cerris*
10	Oak woods with *Potentilla alba*
11	Xero-thermophilous oak woods with *Quercus pubescens* and coline limestone grasslands
12	Oak forests with *Quercus pubescens* and *Fraxinus ornus*
13	Oak forests with *Acer tataricum* and *Quercus pubescens*
14	Acidophilous oak woods
15	Lime-maple woods in lower positions
16	Submountain beech forests
17	Beech and fir-beech forests
18	Calciphilous beech forest
19	Carpathian relicts of pine forests
20	Acidophilous pine woods and sand-dune grasslands
21	Mountain beech forests
22	Mountain maple woods
23	Fir woods and fir-spruce woods
24	Spruce forests with *Vaccinium myrtillus*
25	Spruce forests with tall-herb undergrowth
26	Spruce waterlogged forests, spruce bogs
27	Spruce-pine forests and mountain limestone grasslands
28	Subalpine mountain formations with *Pinus mugo* on calcareous substrates
29	Subalpine mountain formations with *Pinus mugo* on acid substrates
30	Alpine acidophilous grasslands
31	Alpine calciphilous grasslands
32	Raised bogs and transition mires
33	Rooted floating-leaf water plant communities

4.2.1 Sub-complex CLS—Current Landscape Structure: Elements of Current Landscape Structure as Land Cover

Identification, characteristics and delimitation of land cover (LC), as the material expression of natural and socio-economic processes that are formed by the human land use, is the primary and essential condition for the analysis of land use, causes and consequences of use, assessment of human impact on the landscape, as well as solutions to the problem of ecological stability (Feranec and Oťaheľ 1999).

The database of **elements of land cover** processed in vector format was established by synthesis of the relevant layers contained in ZB GIS, forestry mapping outputs and the topographic documentation at BM 1:10,000. Land cover classes were conveniently categorized and content was controlled so as to take account of legitimate spatial linkages between land cover and other elements of the landscape structure (relief, soil, water) so as to ensure logical consistency of data. The vector layer was topologically modified. This database was processed by Esprit, s.r.o. for the SAS, within the project "Modernization and building technical infrastructure for research and development of the Institute of Landscape Ecology, Nitra branch" (Table 4.17).

Table 4.17 Categories of most frequently used land cover elements

Code/no.—LC	Element of land cover
01	Settlement structures
02	Industrial, commercial and transport sites
03	Sites of mining landfill and construction
04	Sites of urban vegetation, sport and recreation
05	Arable land
06	Permaculture
07	Meadows and pastures (small proportion of bushes)
08	Heterogeneous agricultural sites
09	Deciduous forests
10	Mixed forests
11	Coniferous forests
12	Scrubland
13	Transitional forest-scrub
14	Natural meadows
15	Sites with sparse vegetation
16	Rocks
17	Swamps, marsh
18	Water areas

4.2.2 Sub-complex BAC—Biotic-Anthropic Complex: Elements of Current Landscape Structure as Real Vegetation and Habitats

"*Habitat is terrestrial or aquatic areas of a natural or semi-natural character distinguished by geographical, abiotic and biotic characteristics*" (§2 paragraph 2 points) of Act 543/2002 Coll. on the nature and landscape conservation, as amended later.

Forest habitats are forest ecosystems, i.e. forest land and non-forest lands with forest plants, of course, bound with other abiotic elements of a geo/ecosystem and plant and animal species. These are ecosystems in which the ligneous component dominates. In Slovakia they represent the most natural vegetation communities, sometimes similar to climax stage. They are important and dominant component of the landscape, in which they carry out many ecological and socio-economic functions.

Information on forest and meadow habitats can be downloaded from the Information System of taxons and habitats (ISTB). It is a database system designed for the collection and processing of botanical, zoological and habitat data, which plays the role of a national database under Act 543/2002 Coll. on nature and landscape conservation, implementing the regulation of the MoE SR 24/2004 Coll. and to other acts, as amended. ISTB is part of the information system of State Nature Conservation of the Slovak Republic (SNC SR).

Data on forest habitats can also be taken from the existing forestry databases, for example from the Forestry Geographic Information System (LGIS) operated by National Forestry Centrum (NFC) in Zvolen and functionality available to users via the Internet, as a comprehensive information tool for science and research, economic and public practice.

It is also possible to use other available information systems focused on the flora, fauna and habitats, such as

- central database of phytocenological records, which gathers data available of a phytocenological nature from Slovakia. It was begun by the department geobotany at the Institute of Botany at SAS;
- information system on non-forest habitats—grasslands and peatlands, which is operated by the Daphne Institute of Applied Ecology. As part of nationwide mapping it has mapped more than 17,000 sites and is the largest information system on non-forest habitats in Slovakia having charted more than 93% of the total area of grasslands in Slovakia.

Information on forest habitats in our database was taken from an existing Forest Management Plan (FMP) (recently Forest Care Programme—FCP) database (National Forest Centre—NFC) with recoding of forest typology units on forest habitats Schwarz et al. (In Stanová and Valachovič (eds) 2002) (Table 4.18).

Non-forest habitats are specific ecosystems and communities, within which for various reasons (natural and anthropogenic) there is not as dominant ligneous component. Non-forest communities are an equally important part of the landscape. They

Table 4.18 Categories of forest habitats (types of real forest vegetation)

Code/no.—FH	Forest habitats
01	Mixed ash-alder alluvial forests of temperate and Boreal Europe (*Alno-Padion, Alnion incanae, Salicion albae*); White willow gallery forests
02	Riparian mixed forests of *Quercus robur, Ulmus laevis* and *Ulmus minor, Fraxinus excelsior* or *Fraxinus angustifolia*, along the great rivers of the Middle-European provinces (*Ulmenion minoris*); Mixed oak-elm-ash forests of great rivers
03	Mixed ash-alder alluvial forests of temperate and Boreal Europe (*Alno-Padion, Alnion incanae, Salicion albae*); Medio-European stream ash-alder woods
04	Mixed ash-alder alluvial forests of temperate and Boreal Europe (*Alno-Padion, Alnion incanae, Salicion albae*); Montane grey alder galleries
05	Eastern oak-hornbeam forests—Carpathian
06	Eastern oak-hornbeam forests—sub-Pannonian
07	Mixed lime-oak-hornbeam forests—sub-continental
08	Pannonian woods with *Quercus pubescens;* Pannonian (*Quercus pubescens*) woods
09	Euro-Siberian steppe oak woods; Pannonic steppe oak woods
10	Euro-Siberian steppe oak woods; Western white cinquefoil sessile oak woods
11	Pannonian-Balkanic turkey oaksessile oak forests; Pre-Carpathian *Quercus cerris–Quercus petraea* forests
12	Acidophilous, thermophilous and supra-mediterranean oak forests
13	Old acidophilous oak woods with *Quercus robur* on sandy plains
14	*Tilio-Acerion* forests on slopes, screes and ravines
15	*Asperulo-Fagetum* neutrophilous beech forests
16	*Luzulo-Fagetum* acidophilous beech forest
17	Medio-European subalpine beech woods (with *Acer* and *Rumex arifolius*)
18	Medio-European limestone beech forests (*Cephalanthero-Fagion*)
19	Sub-continental acidophilous Scots pine forests
20	Western Carpathian calcicolous *Pinus sylvestris* forests
21	Carpathian steppe pine woods
22	Bog woodland
23	Alder swamp woods
24	Fir and fir-spruce forests
25	Acidophilous spruce forests (*Vaccinio-Piceetea*); Alpine and Carpathian subalpine spruce forests
26	Acidophilous spruce herbal forests (*Vaccinio-Piceetea*); Alpine and Carpathian subalpine spruce herbal forests
27	Acidophilous spruce forests (*Vaccinio-Piceetea*); Alpine and Carpathian subalpine spruce forests, inner range montane spruce forests

(continued)

Table 4.18 (continued)

Code/no.—FH	Forest habitats
28	Alpine *Larix decidua* and *Pinus cembra* forests
29	Bushes with *Pinus mugo*; Dwarf mountain pine scrub
30	Pannonic inland sand dune thicket (*Junipero-Populetum albae*)

occur both naturally in areas where there is no forest due to climate or soil conditions in the climax community, but to a much greater extent in response to anthropogenic changes in real vegetation and creating the cultural steppe. Despite a smaller spatial complexity of non-forest communities and the mainly anthropogenic origin of most of the existing non-forest habitats, these are a very important landscape element and contain a significant proportion of biodiversity.

Information on non-forest habitats (grassland habitats and peat) were taken from the information system on non-forest habitats (Daphne Institute for Applied Ecology) (Table 4.19).

4.3 Indicators of Properties of Tertiary Landscape Structure

The tertiary structure of the landscape is characterized by data on socio-economic phenomena (factors)—SEF. Their spatial manifestation has the character of areas, sections, bands, zones, sites, regions, and protected areas. They are defined by the rules of a legal nature—in acts, regulations, directives, standards, codes of practice, conventions, as well as documents of a conceptual nature such as plans, projects, further in development documents such as the Programme of economic and social development of communities, local Agenda 21, in documents of territorial systems of ecological stability, in sectoral development concepts and so on. Spatial plans and projects are actually also a set of expressions of complex socio-economic phenomena. Until the time of their execution they are intangible regulations, they express the interests of man on land use. SEF carry the **guidelines, restrictions and prohibitions** to human activities are therefore very important factor in the care of the environment (Miklós and Špinerová 2011).

Socio-economic phenomena are **linked to elements of primary and secondary landscape structures**. There may be a large number of them in the landscape depending on the level of detail of the processed area. An overview of the most commonly mapped socio-economic phenomena at scales of 1:10,000 to 1:100,000, as well as their relationship to the elements of primary and secondary landscape structures are given in Tables 4.20, 4.21, 4.22 and 4.23 (see below) . In this selection are only the SEFs, which are defined in any law or in other regulations, guidelines, methodologies and so on.

Table 4.19 Categories of non-forest habitats (types of real non-forest vegetation)

Code/no.—NFH	Groups of non-forest habitats
01	**Marshes and habitats with the occurrence of halophytes**: Inland salt meadows; Pannonic salt steppes and salt marshes; Pannonic saline meadows
02	**Sands and pioneer vegetation**: Pannonic inland dunes; Xeric sand calcareous grasslands; Dwarf annual siliceous grasslands; Siliceous rock with pioneer vegetation of the *Sedo-Scleranthion* or of the *Sedo albi-Veronicion dillenii*; Rupicolous calcareous or basophilic grasslands of the *Alysso-Sedion albi*
03	**Aquatic habitats**: Oligotrophic to mesotrophic standing waters of plains to subalpine levels with vegetation belonging to *Littorelletea uniflorae* and/or *Isoëto–Nanojuncetea*; Natural eutrophic lakes with *Magnopotamion* or *Hydrocharition*—type vegetation; Natural dystrophic lakes and ponds; Water courses of plain to montane levels with the *Ranunculion fluitantis* and *Callitricho-Batrachion* vegetation; Hard oligo-mesotrophic waters with benthic vegetation of *Chara* formations; Industrial lagoons and ornamental ponds, Eutrophic waters; Shallow-water floating communities; Medium tall waterside communities; Small galingale swards
04	**Non-forest riparian vegetation**: Unvegetated river gravel banks; Alpine rivers and the herbaceous vegetation along their banks; Alpine rivers and their ligneous vegetation with *Myricaria germanica*; Alpine rivers and their ligneous vegetation with *Salix elaeagnos*; Muddy river banks with *Chenopodion rubri* p. p. and *Bidention* p. p. vegetation; Hygrophilous tall herb fringe communities of plains and of the montane to alpine belts; Small reed beds on fast flowing waters
05	**Scrubland and bush habitats**: European dry heaths; *Juniperus communis* formations on heaths or calcareous grasslands; Sub-Arctic willow scrub; Central European sub-Mediterranean deciduous thickets; Medio-European rich-soil thickets; Mire willow scrub; Almond willow-osier scrub
06	**Alpine vegetation**: Siliceous alpine and boreal grasslands; Alpine and subalpine calcareous grasslands; Hygrophilous tall herb fringe communities of plains and of the montane to alpine belts; Alpigene tall grass communities; Alpine and boreal heaths
07	**Thermo and xerophillous and grass-herbal vegetation**: Semi-natural dry grasslands and scrubland facies on calcareous substrates (*Festuco-Brometalia*)—important orchid sites; Sub-Pannonic steppic grasslands; Pannonic loess steppic grasslands; Pannonic sand steppes; Rupicolous pannonic grasslands (*Stipo-Festucetalia pallentis*); Xero-thermophile fringes; Mesophile fringes; Species-rich Nardus grasslands, on siliceous substrates in mountain areas and submountain areas in continental Europe
08	**Meadows and pastures**: Lowland hay meadows (*Alopecurus pratensis, Sanguisorba officinalis*); Mountain hay meadows; Mesophile pastures; *Molinia* meadows on calcareous, peaty or clayey-silt-laden soils (*Molinion caeruleae*); Hygrophilous tall-herb fringe communities of plains and of the montane to alpine belts; Eutrophic humid grasslands; Alluvial meadows of river valleys of the *Cnidion dubii* alliance; Tall rush swamps; *Phalaris arundinacea* beds and large *Carex* beds; Common reed beds; Halophile clubrush beds

(continued)

Table 4.19 (continued)

Code/no.—NFH	Groups of non-forest habitats
09	**Bogs and marshes**: Active raised bogs; Degraded raised bogs (still capable of natural regeneration); Transition mires and quaking bogs; Depressions on peat substrates of the *Rhynchosporion*; Calcareous fens with *Cladium mariscus* and species of the *Caricion davallianae*; Alkaline fens; Transitional tall herb humid meadows
10	**Springs**: Soft water bryophyte springs; Bittercress springs; Petrifying springs with tufa formation (*Cratoneurion*);
11	**Bog and debris habitats**: Chasmophytic vegetation of calcareous rocky slope; Chasmophytic vegetation on siliceous rocky slopes; Siliceous scree of the montane to snow levels (*Androsacetalia alpinae* and *Galeopsietalia ladani*); Calcareous and calcshist screes of the montane to alpine levels—*Thlaspietalia rotundifolii*; Medio-European siliceous scree; Medio-European calcareous scree of hill and montane levels; Caves not open to public
12	**Ruderal habitats**: Herbaceous clearings; Shrubby clearings; Ruderal communities; Field margin cropland, Extensive cultivation; Improved grasslands, Crops; Watercourse veils; Plantation; Industrial lagoons and reservoirs, canal

Depending on the character of SEF they spatial structure can be interpreted in two systems, namely:

(a) as part of the ***territorial system of ecological stability (TSES)***

Into the system we include:

- **SEF of nature conservation and SEF of protection of natural resources** (Tables 4.20 and 4.21), see below . The connection in series and the overlapping structure is strengthened by mutually supportive SEF especially in the field of conservation and protection of water resources, in many places, therefore, it is possible to speak of the multifunctional protection of nature and natural resources. However, there are also opposite situations, for example SEF of commercial forest resources is contrary to the interests of nature conservation. For example, the Act 326/2005 Coll. on Forests forces the owners and users to remove the fallen trees from wind and other natural calamities in each such case. Act 543/2002 Coll. on nature and landscape conservation, as amended, denies any intervention in the areas with the highest degree of protection. Generally, however, SEF of this group from a landscape-ecological perspective, are ***positive factors/phenomena*** (Izakovičová et al. 2000), might be considered as a priority of landscape management (Hrnčiarová et al. 1997).

These SEF act as regulating and limiting realization criteria for other required or proposed human activities.

The material carriers of this structure are elements of primary and secondary landscape structure—specific habitats that within the TSES are defined as *bio-centres, bio-corridors and interactive elements* forming the skeleton of TSES, as well as other elements of the primary and secondary landscape structure that are under some protection. We consider the comprehensive projection of SEF of nature conservation and natural resources protection, as well as the tangible elements of primary and secondary landscape structure to which the SEF are bound, to be the TSES in the broadest sense, as a real whole space covering structure in support of maintaining ecological quality of the landscape (Miklós and Hrnčiarová (eds) et al. 2002; Miklós et al. 2011a, b, c).

(b) as part of the ***territorial systems of stress factors (TSSF)***

Into this system we include:

- **SEF of residential, production, transport and technical and community activities** (Table 4.22). They are also non-material but bind to very significant man-made elements of CLS —tangible objects and areas of mining, manufacturing, municipal and other technical activities that can be considered as **primary stress factors**;
- **SEF binding to deterioration phenomena** (Table 4.23) , which in acts, planning and development documents are defined as **areas, territory, sections, declared burdens and pollution sources** depending on the type and size of these deterioration phenomena (Izakovičová et al. 2000; Špinerová 2010; Miklós et al. 2011a, b, c; Miklós and Špinerová 2011, 2013; Izakovičová and Moyzeová 2012);
- **SEF delimiting area threatened with other stress factors**, which have the character of natural processes, but are caused or influenced by human intervention in primary or secondary landscape structure, such as erosion, landslides, avalanches, floods and other environmental burdens (Tables 4.24, 4.25 and 4.26).

In practice, most commonly used SEF of nature conservation and natural resources protection, as well as SEF of stress factors. We can define the following sub-complexes of SEF.

$$\mathbf{SEC} = (\mathbf{NAC}, \mathbf{NAR}, \mathbf{URB}, \mathbf{CAT}, \mathbf{IND}, \mathbf{AGR}, \mathbf{ADM}, \mathbf{DET}),$$

4.3.1 Sub-complex NAC: Socio-economic Factors for Nature and Landscape Conservation

These SEF are defined on the basis of several separate databases, compiled and managed by several institutions. The list comprises information on conservation areas of the national network, on the Natura 2000 network, on the protected areas declared by other international conventions, on the elements of TSES, territories with protection of natural resources, about ecologically important landscape elements without legislative protection, as well as cultural-historically valuable forms of land use (Table 4.20).

Table 4.20 Selected socio-economic factors of nature conservation declared in acts and other planning and development documents

Elements of primary and secondary landscape structure to which the respective SEF are bounded	SEF defined in various regulations
Specific landscape structure of a predominantly natural character	Different categories of nature conservation areas and their buffer zones: • Protected Landscape Area (PLA) • National Park (NP), NP protection zone (PZ NP) • Protected Area (PA), PA protection zone (PZ PA) • Nature Reservation (NR) • National Nature Reservation (NNR), buffer zone NR and NNR • National Natural Monument (NNM) • Nature Monument (NM), caves, natural waterfall • Protection zone of NM and NNM • Private protected areas
	Zones of PLA and NP
	NATURA 2000 • Special areas of conservation—SCI—Sites of Community Importance • Special protection areas—PBA—Protected Bird Areas
	Sites of the UNESCO World Cultural and Natural Heritage
	Biospheric reservation of the UNESCO's Man and Biosphere Programme
	Wetlands by Ramsar convention
	Elements of TSES in development documents, plans and projects • Protected landscape element—biocentre TSES • Protected landscape element—biocorridor TSES • Protected landscape element-interactive TSES element
Significant landscape structures and elements	Historical landscape structures
Important species and trees	Gene pool sites of flora and fauna • Protected plants, animals • Protected trees (PT) and their buffer zones • Protected minerals, fossils/petrifications

4.3.2 *Sub-complex NAR: Socio-Economic Factors of Protection of Natural Resources*

This group is also relatively rich in SEF. The most common ones are shown in Table 4.21.

Table 4.21 Selected socio-economic factors of protection of natural resources declared in acts and other planning and development documents

Elements of primary and secondary landscape structure to which the respective SEF are bounded	SEF defined in various regulations
Forests	Protective forests: • In exceptionally unfavourable places • High mountain forests, forests in the scrub pine belt • Other forests with predominant protection of soil
	Forests of specific designation: • In protective zones of water resources • In protective zones of healing and mineral resources (spa forests) • In nature conservation areas • Forests designated for research • Forests with medical-recreational function, nearby medical institutions • Suburban forests with health-recreational function, forest parks • In recognized animal and pheasant grounds • Forests under the influence of emissions
	Commercial and other forests: • Timber production forestry • Forest nursery • Military forests • Forests land in Sites of Community Importance • Forest land in declared Protected Bird Area
Water	Protected water area (protected area of natural accumulation of water)
	Water-supply rivers and reservoirs
	Catchment of water-supply rivers and reservoirs
	Important water flows
	Zones of hygiene protection of underground and surface water of 1st to 3rd degrees (protective zone of water sources)
	Protective zones of reservoirs

(continued)

Table 4.21 (continued)

Elements of primary and secondary landscape structure to which the respective SEF are bounded	SEF defined in various regulations
	Water suitable for various functions: bathing, irrigation, for fish breeding and reproduction of original species of fish
Waters with specific properties (natural healing sources)	Protective zones 1–3 level for natural healing water sources,
	Natural healing sources of gases and emanations
	Natural healing resources from peat fens, sludges and other earth healing sources
	Protective zones 1–3 level for natural sources of mineral water
	Spa cities (natural healing spa), protective zone of spa cities
	Interior spa territory
	Exterior spa territory
Climatic conditions beneficial for healing	Protective zone of climatic condition beneficial for healing
Agricultural soils	Registered parcel, culture, notes on culture
	Bonited soil-ecologic units (BSEU)
	Registered permaculture: vineyard, orchards, permanent grassland for grazing, growing areas
	Agricultural production areas
	Investment to land (melioration, irrigation)
	Agricultural land in Sites of Community Importance
	less-favoured areas in registered soil blocks
Habitat of fish and animal	Protected fishing areas
	Fishing grounds
	Fish-farming flows
	Recognized animal keepers
	Independent pheasantries
Minerals	Protected area of deposits with reserved minerals
	Protected area of deposit of non-reserved minerals
	Natural rock structures for storing gas
Historic structures and sites	Cultural monuments and national cultural monuments
	Memorial City Reservation (MCR) and Folk Architecture Reservation (FAR)
	Monuments zone—urban and rural
	Archaeological find
	Protective zones of the memorial fund

4.3.3 Sub-complex URB, CAT, IND, AGR: Socio-Economic Factors of Urbanisation, Industry, Agriculture Production, Transport and Communal Activities

SEF of this sub-complex have the character of **protective, hygiene and safety zones** (Table 4.22). They bind to the material elements that are the source of the barrier effect, pollution, noise and other environmental deterioration factors. It should be emphasized that these carriers of SEF—the material elements of CLS of the nature of the primary stress factors—are indeed stress factors, but their protection, hygiene and safety zones **themselves are not negative phenomena!** On the contrary, they have a role to play restricting and regulating function of the impacts of deterioration with other activities and populations. From the territorial point of view, these bands reflect the projected area of direct and strongest **influence of primary stress factors**, therefore these zones defined areas can be considered as adversely affected territory, perhaps with lower environmental quality, as they have areas outside the zones. This group of SEF includes also other **functional areas**, for example urban and recreational areas.

Table 4.22 Selected socio-economic factors of urbanisation, industrial, transport, technical and communal activities declared in laws and other planning and development documents

Elements of primary and secondary landscape structure to which the respective SEF are bounded	SEF defined in various regulations
Residential areas	Residential zones of individual and complex apartment building
	Zones of public services
	Cultural and school zones
	Zones of relaxation and sport, parks and public greenery,
	Protective zones of hospital areas
Recreational areas	Declared recreational area, zones and centres of varying levels
	Declared campgrounds
Production areas	Hygienic protective zones—HPZ/safety zones—SZ of industrial areas areas
	HPZ/SZ of quarries and mining sites
	HPZ/SZ of waste landfills
	HPZ/SZ of wastewater treatment
	HPZ/SZ of agricultural objects
	HPZ/SZ of agrochemical businesses
	HPZ/SZ of animal productions

(continued)

Table 4.22 (continued)

Elements of primary and secondary landscape structure to which the respective SEF are bounded	SEF defined in various regulations
	HPZ/SZ of store of manure and field airports
Transport and other structures	Protective zones of railways
	Protective zones of roadways
	Protective zones of airports and landing path of airports
	Protective zones of ports
	Protective zones of electrical cables
	Protective zones of gas facilities
	Protective zones of gas and oil pipelines
	Protective zones of cables
	Protective zones of storages
	Military zones

4.3.4 Sub-complex DET: Socio-Economic Factors Bound to the Deterioration of the Environment

These SEF are often referred to using the term stress factors in the landscape. As with the previous group, these SEF are also covered by a variety of institutions.

Within this group we distinguish in principle stress factors, which are:

(a) caused by permanent human activity—this is mainly pollution of individual geosystem elements, such as pollution of air, water, soil, biota. Deterioration phenomena are presented in Table 4.23, in particular, pollution and other factors that threaten the elements of the primary and secondary landscape structure, as well as mankind. They have the character of **secondary stress factors** because they occur depending on the primary stressors, and because these phenomena do not always occur, or with the same intensity. E.g. the current degree of pollution of water and air is constantly changing, sometimes it can reach the zero value, but objects that are the source—including primary stressors—are always in the landscape. The SEF of this group delimits the area that is affected by deterioration.

The main carriers of the above SEF are primarily objects and areas of mining, manufacturing, municipal, transportation and other technical activities that have the character of **primary stress factors** (see Table 4.23). Their primary stress manifestation is considered both directly their physical existence in the country, which manifests itself as a barrier effect, as well as the production of pollutants, noise, radiation, vibration, of which these objects are the **source**;

(b) stress factors that occur as direct consequences of previous human activities, for example. various types of old environmental burdens, undermined areas, landslides due to construction activities, as well as gully erosion and other damage to soil;

(c) stress factors of natural hazards that mankind could in the past, to a greater or lesser extent, initiate by intervention in the primary or secondary landscape structure, but are currently under way without his direct intervention, for example avalanches, slope deformations, landslides, erosion, flooding and other environmental burdens (Tables 4.24 and 4.25).

These areas are also defined in acts, maps, and other development documents.

(d) stress factors that occur without the cause of mankind. These can no longer be called socio-economic phenomena, in essence, they are eminent processes of the primary landscape structure, but because they also have stressful manifestation, they can be added to the characteristics of the geosystems here. This special group of this sub-complex forms is called **geodynamic events**.

Table 4.23 Selected socio-economic factors bound to the deterioration of the environment declared in acts and other planning and development documentation

Elements of primary and secondary landscape structure to which the respective SEF are bounded	SEF defined in various regulations
Sources of pollution	Stationary sources of atmospheric pollution • Particularly important technological unit ("large source") • Important technological unit, if not part of a large source ("medium source") • Other technological units ("small source")
	Mobile sources of air pollution
	Sources of water pollution (differed according to size)
	Noise sources (depending on size)
	Sources of radiation
Territory with air pollution	Area of air pollution • Area of air quality management • Defined parts of zones or agglomerations with an area of 50 km^2 • Where there occur odorous pollutants in concentrations which are nuisance to the population
	Area burdened by ground-level inversions

(continued)

Table 4.23 (continued)

Elements of primary and secondary landscape structure to which the respective SEF are bounded	SEF defined in various regulations
	Area burdened by noise (by degree)
Territory with water pollution	Sections of polluted water courses (by category)
	Area of polluted groundwater above the norm
	Sensitive areas
	Vulnerable areas
	Declared inundation area
Soil damage and soil with reduced conditions	Territory with contaminated soil
	Erosion of damaged soil (by degree)
Damaged vegetation	Damage to vegetation (by degree)
	Forests damaged by immisions
Other areas threatened by stress factors	Environmental burden according to Register of environmental burdens A, B, C
	Area threatened by landslides
	Area threatened by avalanches
	Undermined area
	Area of increase seismicity
	Area of increased natural radioactivity

4.3.5 Sub-complex GDP: Geodynamic Phenomena with the Character of Stress Factors

In the national database of abiocomplexes there are essentially distinguished two types of stress factors: stress factors of the character of natural hazards—**geodynamic phenomena** and **anthropogenic stress factors**.

Attribute values have been evaluated according to the data source from different projects implemented since 1963 archived in the SGIDŠ database, for example for the atlas of maps of slope stability, register of landslides, in digital geological maps, engineering-geological zoning, register of slope deformations, neotectonic map of Slovakia.

From the point of view of geosystem, SEF of the natural stress factors listed in Table 4.23 under "Other areas threatened by various stress factors" are recorded or registered to the **administrative** territory where the potential or real phenomena occur. Geodynamic phenomena on the other hand are basically **indicators of properties of abiocomplexes**. Accordingly, the occurrence of geodynamic hazards (threats) or susceptibility of an area to their development is tracked to an abiocomplex, where they actually occur. This information is used mainly for engineering purposes. We could include the following factors:

SD—Slope Deformation

Phenomena are mapped such as massif tearing, massif loosening, block disintegration, block fields, landslides, slope flows, rock collapse, undermined landslides, undermined blocks and combinations thereof.

NA—Neotectonic Activity

Neotectonic activity is closely related to seismicity of the area. Activity in the vicinity of tectonic lines can affect other geodynamic phenomena—such shocks can be the cause—trigger of landslide formation. It is also a potential cause of earthquakes.

Manifestations of neotectonic activities were derived mainly from overview neotectonic maps of Slovakia on a scale of 1:500000 (Maglay et al. 1999). Each abiocomplex is assigned to the database field, one of the two code numbers: 1—neotectonic area with no evidence of activity, 2—neotectonic active area in the turning zone (in the vicinity of the active fault).

GE—Gully Erosion

Gully erosion is characterized by the presence of gully erosion in the abiocomplex. The location of erosion gullies was taken from maps on a scale of 1:50,000. Abiocomplexes in which erosion gullies are not found have in the database field the values: 1—abiocomplexes in which erosion gullies are found and 2—no erosion gullies.

RLS—Re-location of Loess Sediments

The settling of fine grain-size soil can occur if there are any of the following conditions: soil is of aeolian origin, the content of silt component is more than 60% by weight of dry soil, the content of the clay component is less than 15% by weight of dry soil, degree of saturation is less than 60% and the limit of fluidity is less than 32%. Fine-grained soils are prone to settling if their porosity is greater than 40%, while their natural humidity less than 13%. According to the cited standards sediments are settling, if their settling coefficient is higher than 1% in vertical tension corresponding to the weight of the overburden, or overburden weight and load.

KP—Presence of Karst Phenomena

The presence of karst phenomena and rocks susceptible to karst formation (solid limestone) in the abiocomplex is expressed as a code in database fields: 1—area not susceptible to the formation of karst phenomena, 2—area with tendency to form karst phenomena.

WL—Presence of Waste Landfill

Expressed dually: 1—landfill not present in abiocomplex, 2—landfill present in abiocomplex. Also available is information on their size. Information on the landfill of waste listed in the database was created as part of landfills registration all over Slovakia and is managed in the archives at SGIDŠ.

Table 4.24 Types of slope deformations

Code/no.—SD	Type of slope deformation
01	Block disintegration
02	Block disintegration + block fields
03	Block disintegration + block fields + landslides
04	Block disintegration + slope stream flows
05	Block disintegration + rock collapse
06	Block disintegration + rock collapse + massif tearing and massif loosening
07	Block fields
08	Block fields + rock collapse
09	Block fields + landslides
10	Undermined blocks
11	Undermined blocks + block fields
12	Undermined block fields
13	Undermined slope stream flows
14	Undermined rock collapse
15	Undermined landslides
16	Slope stream flows
17	Slope stream flows + rock collapse
18	Tearing and loosening of massif
19	Massif tearing + block disintegration
20	Massif tearing + block disintegration + rock collapse
21	Massif tearing + rock collapse
22	Rock collapse
23	Rock collapse + block disintegration
24	Rock collapse + slope stream flows
25	Rock collapse + massif tearing
26	Rock collapse + massif tearing + block disintegration
27	Landslides
28	Landslides + block fields
29	Landslides + slope stream flows
30	Landslides + rock collapse
31	Slope stream flows + block fields

Table 4.25 Degrees of susceptibility to landslides and to re-location of loess sediments

Code/no.—RLS	Degree of susceptibility to settling
01	Environment is not susceptible to settling
02	Area with soils weakly susceptible to settling
03	Area with soils susceptible to settling
04	Area with soils highly susceptible to settling
0	Area not assessed as it is built of rock or semi-rock stones

4.3.6 Sub-complex ADM: Socio-Economic Factors of the Character of Administrative and Sectoral Boundaries

These are also typical non-material, intangible elements of the tertiary landscape structure, which however has a very significant impact on the landscape management. They create administrative and territorial legal framework for the application of all spatial planning processes. The most important boundaries in this respect are shown in Table 4.26.

Table 4.26 Selected socio-economic factors of the character of administrative and sectoral boundaries declared in acts and other planning and development documents

Elements of primary and secondary landscape structure to which the respective SEF are bounded	SEF defined in different regulations
Holistic landscape units (landscape as a whole)	Administrative territorial division: • State, regional, district, municipal boundaries • Cadastral boundaries, intraurban boundaries • Boundaries of collective and land owner communities • Boundaries of territorial planning units (region, settlement, zone)
Sectorial territorial units	Sectoral boundaries of various character: • Administrative territory of river basin • Boundaries of forestry management units (FMU) • Boundaries of agricultural companies • Boundaries of hunting grounds • Boundaries of fish-farming and sections of fishing grounds

References

Bodiš D, Čurlík J, Liščák P, Marcin D, Pristaš J, Rapant S, Smolárová H, Zakovič M (1998) Súbor regionálnych máp geofaktorov životného prostredia 1:50,000. Archív odboru Geofondu ŠGÚDŠ. arch. č. 82900, Manuskript, 59 pp

Collective (2000) Morfogenetický klasifikačný systém pôd Slovenska. Bazálna referenčná taxonómia. VÚPaOP, SPS, Bratislava, 74 pp

Feranec J, Oťaheľ J (1999) Mapovanie krajinnej pokrývky metódou CORINE v mierke 1:50,000: Návrh legendy pre krajiny programu Phare. Geografický časopis, 51:19–44

Hraško J, Linkeš V, Němeček J, Novák P, Šály R, Šurina B (1991) Morfogenetický klasifikačný systém pôd ČSFR. Bazálna referenčná taxonómia. Referenčný klasifikačný systém. Klasifikácia pôdotvorných substrátov. 2. doplnené vydanie, VÚPÚ, Bratislava, 106 pp

Hrnčiarová T et al (1997) Ekologická únosnosť krajiny: metodika a aplikácia na 3 benefičné územia. Časť I. – IV. Ekologický projekt, MŽP SR, Bratislava, ÚKE SAV, Bratislava, Svetová banka GEF, 490 pp

Izakovičová Z, Moyzeová M (2012) An example of transport of the scientific knowledge in the real practice. In: Forum Carpaticum 2012: from Data to Knowledge, from Knowledge to Action [elektronický zdroj]. Institute of Landscape Ecology, Slovak Academy of Sciences, Nitra, pp 114

Izakovičová Z, Hrnčiarová T, Králik J, Liška M, Miklós L, Moyzeová M, Pauditšová E, Ružičková H, Šíbl J, Tremboš P (2000) Metodické pokyny na vypracovanie projektov regionálnych územných systémov ekologickej stability a miestnych územných systémov ekologickej stability. MŽP SR, Združenie KRAJINA 21, Bratislava, 155 pp

Kobza J, Barančíková G, Čepková V, Došeková A, Fulajtár E, Houšková B, Makovníková J, Matúšková L, Medveď M, Schlosserová J, Styk J, Vojtáš J (2002) Monitoring pôd Slovenskej republiky súčasný stav a vývoj monitorovaných vlastností pôd. Výsledky čiastkového monitorovacieho systému – pôda, ako súčasť Monitoringu životného prostredia Slovenskej republiky za obdobie 1997–2001. SSCRI Bratislava, 180 pp

Krásný J (ed), Daňková H, Hanzel V, Kněžek M, Matuška M, Šuba J (1981) Mapa odtoku podzemní vody ČSSR. Český hydrometeorologický ústav, Kartografie, n.p. Praha. Mapa 1:1,000,000

Linkeš V, Gromová A, Lupták D, Poliak P (1988) Informačný systém o pôde. Príroda, Bratislava

Linkeš V, Kobza J, Švec M, Ilka P, Pavlenda P, Barančíková G, Matúšková L et al. (1997) Monitoring pôd SR. Súčasný stav monitorovaných vlastností pôd 1992–96. VÚPÚ Bratislava, ÚKSUP Bratislava, Lesoprojekt Zvolen, 128 pp

Maglay J et al (1999) Neotektonická mapa Slovenska 1:500,000. Ministerstvo životného prostredia Slovenskej republiky. Geologická služba Slovenskej republiky, Bratislava

Malík P, Bačová N, Hronček S, Ivanič B, Káčer Š, Kočický D, Maglay J, Marsina K, Ondrášik M, Šefčík P, Černák R, Švasta J, Lexa J (2007) Zostavovanie geologických máp v mierke 1:50,000 pre potreby integrovaného manažmentu krajiny. ŠGÚDŠ, Manuskript – archív Geofondu, Bratislava, 552 pp

Malík P, Bahnová N, Ivanič B, Kočický D, Maretta M, Špilárová I, Švasta J, Zvara I (2011) Komplexná geologická informačná báza pre potreby ochrany prírody a manažmentu krajiny. Záverečná správa + prílohy. MŽP SR, ŠGÚDŠ, Bratislava, 149 pp + prílohy

Mazúr E, Činčura J, Kvitkovič J (1980) Geomorfologické pomery. Mapa v mierke 1:5,00,000

Michalko J et al (1986) Geobotanická mapa ČSSR. Slovenská republika./Textová časť/, Veda, Bratislava, 168 pp + 40 pp prílohy

Miklós L, Hrnčiarová T (eds) et al (2002) Atlas krajiny Slovenskej republiky. 1. vydanie, MŽP SR Bratislava, SAŽP Banská Bystrica, 344 pp

Miklós L, Izakovičová Z (1997) Krajina ako geosystém. VEDA, SAV, Bratislava, p 152

Miklós L, Špinerová A (2011) Krajinno-ekologické plánovanie LANDEP. VKÚ, Harmanec, p 159

Miklós L, Špinerová A (2013) Krajinné plánovanie. Praktikum. Technická univerzita Zvolen, 96 pp

Miklós L, Diviaková A, Izakovičová Z (2011a) Ekologické siete a územný systém ekologickej stability. Vydavateľstvo TU vo Zvolene, 141 pp

Miklós L, Ivanič B, Kočický D (2011b) Krajinnoekologická základňa integrovaného manažmentu povodia Ipľa. Digitálna databáza a tematické mapové vrstvy. Projekt HUSK 0801/2.1.2/0162 Vytvorenie jednotného monitoringu na báze priestorového informačného systému v povodí Ipľa. Banská Štiavnica, Esprit, spol. s r. o. (elektronický zdroj)
Miklós L, Izakovičová Z, Kanka R, Ivanič B, Kočický D, Špinerová A, David S, Piscová V, Štefunková D, Oszlányi J, Ábrahámová A (2011c) Geografický informačný systém povodia Ipľa: Katalóg GIS a výber máp. Bratislava: Ústav krajinnej ekológie SAV: Katedra UNESCO, Fakulta ekológie a environmentalistiky, Technická univerzita Zvolen, Esprit Banská Štiavnica, 143 pp
Rapant S, Vrana K, Bodiš D (1996) Geochemický atlas Slovenskej republiky – časť I.: Podzemné vody. GSSR Bratislava, 127 pp
Sitková Z (2000) Metodické aspekty priestorového rozloženia klimatických prvkov v horskej oblasti. Bioklimatológia a životné prostredie. XIII. Bioklimatická konferencia SBkS a ČBkS, Košice, 10 p
Skalský R, Balkovič J (2002) Digital database of selected soil profiles of the complete agricultural soil survey of slovakia (KPP-DB). In: Proceedings of SSCRI, SSCRI, Bratislava, 25:129–140
Špinerová A (2010) Krajinno-ekologické limity poľnohospodárskeho využitia povodia Ilijského potoka. VKÚ, a. s, Harmanec, p 118
Stanová V, Valachovič M (eds) et al (2002) Katalóg biotopov slovenska. DAPHNE – Inštitút aplikovanej ekológie, Bratislava, 225 pp
Tarboton DG (1997) A new method for the determination of flow directions and upslope areas in grid digital elevation models. Water Resour Res 33(2):309–319
Thornthwaite CW (1984) An approach toward a rational classification of climate. Geograph Rew 38:55–94
Thornton PE, Running SW, White MA (1997) Generating surfaces of daily meteorological variables over large regions of complex terrain. J Hydrol 190:214–251
Tveito OE (2002) Spatial distribution of winter temperatures in Norway related to topography and large-scale atmospheric circulation. Predictions in Ungauged Basins: PUB Kick-off (Proceedings of the PUB Kick-off meeting held in Brasilia, 20–22 November 2002). IAHS Publications 309, 2007
Water Management Map of the Slovak Republic 1:50000. WRI, Bratislava

Chapter 5
Application of the Spatial Information System of Geocomplexes in Model Territory

Abstract The chapter presents the application of the geosystem approach to the creation of a proper spatial informational system with respects to the principles described in previous chapters. The application is presented on concrete model territory of the watershed of the Ipel' river (South Central Slovakia). Those principles were the followings

- a unified mathematical depiction of the map base, namely in depiction *UTM Zone 34N*; Gauss–Krüger projection of the cartographic base
- the spatial projection base is a digital elevation model of DTM relief;
- the framework for each thematic layer is unified topographic object base. Maps are processed in a uniform format in the *ESRI filegeodatabase*;
- the carriers of spatial information are carefully compiled in the system of georeferencing elements, namely:
  - raster—for morphometric parameters of relief;
  - point—hydrological and weather stations;
  - section—sections of rivers and roads;
  - polygon—indicators of other geosystem elements that are displayed in the form of areals, i.e. abiotic, biotic and socio-economic elements and complexes, including elements of land use and buildings in GIS base.
- contains a purposefully selected set of indicators and values that characterize the primary structure (abiotic complex), secondary structure (biotic complex and land use), as well as the tertiary landscape structure (socio-economic complex), including statistical indicators, which are linked to the village and cadastral area;
- the system of georeferenced elements and indicators enables permanent updating of values as well as the addition of new variables into the system.

The indicators and their domain values were arranged to a purpose-oriented catalogue that allows a quick and easy access to the database and map content. The catalogue was constructed hierarchically, it is possible to gradually gain an overview of the content by gradually unpacking the various hierarchical levels and layers. These hierarchical levels are

L. Miklós et al., *Landscape as a Geosystem*,
https://doi.org/10.1007/978-3-319-94024-3_5

➢ **element** of the geosystem (level of Component);
  ➢ **property** of element (level of Thematic layer);
    ➢ **indicator** of property of element (level of Attribute, indicator);
      ➢ **dimension** of indicator of property of element (level of Dimension);
        ➢ **value** of indicator of property of the element (level of Value of attribute).

As the result the chapter brings as examples cuts of the electronically produced maps on different indicators.

Keywords Model territory · Information system · Catalogue · Maps

All theoretical and methodological provisions on the creation, mapping and content of geocomplexes for the needs of planning processes, which we described in previous chapters, as well as methodical provisions for creation of databases and spatial information system we applied to a number of model areas. This chapter presents an example of construction and mapping of geocomplexes in the basin of the river Ipel' (Slovak Republic), as well as creation of spatial information system.

Application shows the suitability of spatial database in the form of geocomplexes for the practical use.

5.1 Characteristic of Geocomplexes in the Model Territory of the Ipel' Basin

This work was carried out within the project HUSK 0801/2.1.2/0162. The GIS database was performed by ESPRIT, s.r.o. of Banská Štiavnica. Part of the database was processed by the Institute of Landscape Ecology at the SAS in Bratislava and by Közép-Duna völgyi Környezetvédelmi, Természetvédelmi és Vízügyi Felügyelőség (The Inspection for Environment, Nature Conservation and Waters of Middle Danube Valley) Budapest and Cholnoky Jenő Környezetgazdálkodási Dokumentációs és Kutatási Központ Nonprofit Kft (Cholnoky Jenő Environmental, Documentary and Research Centre Nonprofit Ltd.), Budapest. The expanded form of the results of this project is presented in the papers and books of Verrasztó (2010); Miklós and Izakovičová et al. (2011); Miklós et al. (2014).

The aim of this work was the creation of such a spatial information system that respects the following principles of the geosystem approach to landscape:

- Is based on a unified mathematical depiction of the map base, namely in depiction *UTM Zone 34N*; Gauss–Krüger projection of the cartographic base
- Its spatial projection base is a digital elevation model of DTM relief;
- The framework for each thematic layer is unified topographic object base. Maps are processed in a uniform format in the *ESRI filegeodatabase*;
- The carriers of spatial information are carefully compiled in the system of georeference elements, namely:

- raster—for morphometric parameters of relief;
- point—hydrological and weather stations;
- section—sections of rivers and roads;
- polygon—indicators of other geosystem elements that are displayed in the form of sites, i.e. abiotic, biotic and socio-economic elements of geosystems, including elements of land use and buildings in ZB GIS (base for GIS);

- It contains a purposefully selected set of indicators and values that characterize the initial structure (abiotic complex), secondary structure (biotic complex and land use), as well as the tertiary landscape structure (socio-economic complex), including statistical indicators, which are linked to the village and cadastral area;
- The system of georeferenced elements and indicators enables permanent updating of values as well as the addition of new variables into the system (Németh 2016; Kocsis et al. 2016; Miklós et al. 2011a, b).

5.1.1 Building and Hierarchical Arrangement of Databases of Geocomplexes

According to the principles set out in the previous chapters, and according to the objectives of the work and nature of the area, the basic mapping task and creation of the factual content of the database on geocomplexes in the model area is shown as a diagram in Fig. 5.1.

One tool for overview of the database is a special purpose-oriented catalogue that allows a quick and easy access to the plain contents of the database and map

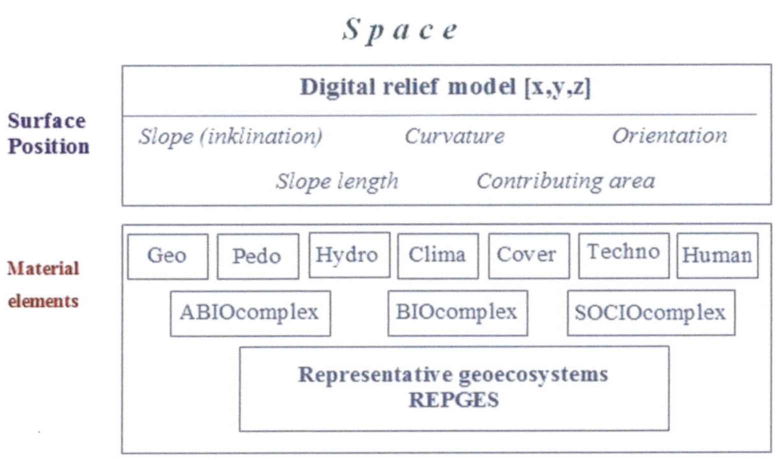

Fig. 5.1 Schema of groups of indicators of properties of elements of geocomplexes for GIS of the Ipel' basin

content. The product was prepared in printed and electronic forms. The catalogue was constructed hierarchically, it is possible to gradually gain an overview of the content of the various hierarchical degrees by gradually unpacking the various hierarchical levels and layers. The final-level catalogue also appears in its own sidebar on GIS thematic maps, including specific values for the indicators of properties of elements of geocomplexes.

The content and hierarchy of the information system is well shown by the comprehensive header of table, which represents the actual catalogue of information layers:

Component	Geo ref. element	Thematic layer	Attribute, indicator	Dimension	Values of attribute or characteristic	Description

The catalogue is compiled based on the logic of the geosystem approach to landscape, namely:

Hierarchical levels *have the following sequence:*

➢ **element** of the geosystem (level of Component);

➢ **property** of element of the geosystem (level of Thematic layer);

➢ **indicator** of property of element of the geosystem (level of Attribute, indicator);

➢ **dimension** of indicator of property of element of geosystem (level of Dimension);

➢ **value** of indicator of element of geosystem (level of Value of attribute).

The second column labelled as "Georeferencing element" has no hierarchical value, it is information on the geoinformation format of the given indicator (point, section, raster, polygon). The last column may contain notes or other descriptions.

Explanation of these levels

- **component** of the landscape—in the geosystematic interpretation, they are the basic elements of the system, which carry its material essence;
- the types of the **georeferential elements** are given for each thematic overlay. In this case, they are vector data—polygons, lines (sections), points—and raster elements;
- **thematic l**ayer—this are the **properties** of the component processed and projected on a map related to a specific database.

Of special importance is the layer KEK_Sk—Landscape-ecological complexes (see later), in which each spatial spot carries the values of 19 properties of the landscape. They are all spatially harmonized according to their real functional relationships. The Ipoly GIS contains 77 thematic layers, to which specific maps are related.

- **attribute/indicator**—each thematic map is regularly described by several indicators;
- **dimension**—it is measure of unit, also given for the value of each indicator. This can be a text (naming), code, date, real numerical data, serial number, ratio, degree;
- **value** of the indicator—this is the most detailed level of the system. In fact, this level represents the database. They contain the values of indicators in the study area. The values can also be seen when clicking on the element.

The catalogue is constructed so that in electronic form the various hierarchical levels can be "unpacked", wrapped in and out, from the most general level to the deepest level, to a list of values of the indicators.

The geosystem components are compiled as follows:

- **Relief**: as a non-material element of landscape and spatial framework of geocomplexes;
- elements of the **primary landscape structure**: abiotic components;
- elements of the **secondary landscape structure**: plants, animals, current land cover,
- **landscape-ecological complex**: synthesis of the elements of the primary landscape structure and land cover;
- elements of the **tertiary landscape structure**: socioeconomic phenomena in the landscape, indicators of environment, indicators pertaining to administrative units—cadastral territory, municipality, region—land area, population, infrastructure, environment.

Due to the breadth of the attributes of the component Current land cover, which is expressed in the thematic layer ***RL001_Sk—Land cover***, (3 hierarchy levels, 14 element groups, 32 elements and 96 parts of elements), the attributes for this component have been elaborated by the specific cataloguing table from which is more clearly to get an overview of the hierarchical order and mutual relations of the three hierarchical levels of elements of the current landscape structure, namely: group of elements, elements and details of elements of the current landscape structure. In addition, for each detailed element of the system, a more detailed characteristic is available in the database, which can be accessed by clicking the information icon on the object (Collective … 2008; Conception … 2006).

Due to the large number and extensive nature of the indicators in the database of thematic layers ***TU001_Sk—Cadastral area*** to which the demographic, social and other data are attached, they are not directly catalogued in this catalogue, but are available through the list of municipalities.

Table 5.1 Overview of information layers at the " Component" level

Component
Relief
Geological Substrate
Soil
Water
Climate
Flora
Fauna
Current land cover
Landscape-ecological complex
Socioeconomic phenomena of protection of nature and monuments
Socioeconomic phenomena of protection of natural resources
Socioeconomic phenomena related to infrastructure
Environment
Territorial units

5.1.2 Content of Information Layers of the Database

As we stated, the database is built on five hierarchical layers, in the order component (element of geocomplex)—thematic layer (property of element)—attribute (indicator of property)—dimension value of attribute—attribute value (value of indicator of property). Individual ranks shown in the table are expandable and foldable.

The content of individual hierarchical layers is presented in the following tables: (Tables 5.1, 5.2 and 5.3)

The thematic layer **KEK_Sk**—Landscape-ecological complex has a specific synthetic character. These are polygons with homogeneous values of 19 indicators of properties of elements of landscape-ecological complex. The list of these indicators is shown in Table 5.4.

The database contains a total of 77 basic thematic layers and more than 500 attributes—characteristics of properties. The complete table is therefore large in size, it is processed electronically, we do not consider it rational to publish it in print. To illustrate a cross-section of the table, we present Table 5.6.

For a large range of indicators, the properties of CLS element, the list of these elements and their indicators are listed in separately in Table 5.7.

Indicators relating to the georeference element, Cadastral areas are listed in separate Table 5.8.

Important note: This book does not deal with the methods of calculating the coefficients, ranks and special values that describe the various functional properties of the abiotic and biotic components. Other specialized institutions and scientists deal with them. Our task was to enter these special values into the GIS and relate them to the proper components and to their properties so that they be also spatially related to the georeferenced elements.

Table 5.2 Overview of information layers at the levels "Component" and "Thematic layer" Primary landscape structure

Component	Type	**Thematic layer—code** and explanation
Relief	Raster	**Hillshade**—shade of relief
	Raster	**Heights**—digital terrain model
	Raster	**Heights_smt**—digital terrain model (smoothed)
	Raster	**Slope**—slope
	Raster	**Aspect**—orientation of relief
	Raster	**Curv_profile**—normal curvature
	Raster	**Curv_plan**—horizontal curvature
	Raster	**Flowlength**—length of slope
	Raster	**Flowacc_d8**—contributing area, microcatchment (D8)
	Raster	**Flowacc_dinf**—contributing area, microcatchment (D-infinite)
	Raster	**Radiation**—sun power
	Raster	**Radiation_dur**—period of sunlight
	Line	**CA010_Sk**—contour
	Point	**CA030_Sk**—spot height
	Polygon	7 indicators of georelief are synthetically projected on layer KEK_Sk—landscape-ecological complex
Geological substrate	Polygon	**PG001_Sk**—site of geological unit
	Polygon	6 indicators of the geological substrate are synthetically projected in the layer KEK_Sk—landscape-ecological complex
Soil	Polygon	**PP001_Sk**—site of soil unit
	Polygon	3 specific indicators of soil are projected in layer KEK_Sk—landscape-ecological complex
Water	Polygon	**PW001_Sk**—detailed catchment
	Polygon	**PW002_Sk**—basic basin
	Polygon	**PW003_Sk**—partial basin
	Polygon	**PW004_Sk**—main basin
	Line	**PW005_Sk**—enclosure
	Line	**RW001d_Sk**—water course
	Line	**RW001_Sk**—line segment of water course
	Polygon	**RW002_Sk**—water area
	Polygon	**SA010_Sk**—area of water flow
	Point	**RM002_Sk**—hydrological station
Air	Point	**RM004_Sk**—meteorological station
	Polygon	**PO002_Sk**—climate geographical area

Table 5.3 Overview of information layers at the levels "Component" and "Thematic layer" Secondary landscape structure

Component	Type	**Thematic layer—code** and explanation
Flora	Polygon	**Fytogeogclen_Sk**—phytogeographical-vegetation classification
	Polygon	**Potveget_Sk**—potential vegetation
	Point	**Botinvdr_Sk**—extension of selected invasive species of plants
Fauna	Polygon	**Zoogeogter_Sk**—zoogeogr. Division—terrestrial biocycle
	Polygon	**Zoogeoglim_Sk**—zoogeogr. Division—limnic biocycle
	Point	**Zoocic_Sk**—extension of selected species of mammals
	Point	**Zoocicvydra_Sk**—extension of species of Eurasian otter (*Lutra Lutra*, Linnaeus, 1758)
	Point	**Zoovtaky_Sk**—extension of selected bird species
	Point	**Zooplazy_Sk**—extension of selected reptile species
	Point	**Zooobojziv_Sk**—extension of selected amphibian species
	Point	**Zoochrobac_Sk**—extension of selected beetle species
	Point	**Zoovazky_Sk**—extension of selected dragonfly species
Current landscape cover	Polygon	**RL001_Sk**—elements of lands cover
Landscape-ecological complex	Polygon	**KEK_Sk**—landscape-ecological complex

Table 5.4 Content of thematic layer **KEK_Sk**—Landscape-ecological complex

Component	Type	**Thematic layer—code** and name	**Attribute, indicator—code** and explanation	**Dimension description**
Landscape-ecological complex	Polygon	**KEK_Sk**—landscape-ecological complex	**NM_VYSKA**—average altitude m a.s.l. of LEC	[m asl]
		Relief	**SKLON**—category of average slope of LEC	Text
			MORFO_POL—morphographic-position type of relief of LEC	Text
			EXPOZICIA—predominant orientation of LEC	Text
			N_FORMA—normal shape of relief of LEC	Text
			H_FORMA—horizontal shape of relief LEC	Text

(continued)

Table 5.4 (continued)

Component	Type	**Thematic layer—code** and name	**Attribute, indicator—code** and explanation	**Dimension description**
		Geological substrate	**GSK**—type of geological-substrate-hydrological complex	Text
			HG_INDEX—hydrogeological types (HGT) of sediments, weatherings and rocks	Text
			HG_T—coefficient of transmissivity of HGT sediments, weatherings and rocks.	[m^2 s^{-1}]
			HG_K—coefficient of filtration of HGT sediments, weatherings and rocks.	[m s^{-1}]
			HG_SV—free capacity of HGT sediments, weatherings and rocks	Number
		Soils	**PODNY_TYP**—soil subtype	Text
			PODNY_DRUH—soil texture (grain size)	Text
			HLBKA_PODY—depth of soil	Text
			SKELET—skeletality of soil	[%]
		Land use	**LANDUSE**—dominant use of LEC	Text
		Landscape-ecological complex	**R**—retention ability of territory	Number
			ER_KAT—erosion threat	Text
			Shape_Area—size of area of LEC	[m^2]
			Shape_Length—circumference of area of LEC	[m]

Table 5.5 Overview of information layers on the levels "Component" and "Thematic layers" for Tertiary landscape structure

Component	Type	**Thematic layer—CODE in system** and explanation
SEF of protection of nature	Polygon	**TK001_Sk**—large protected area
	Polygon	**TK002_Sk**—small protected area
	Polygon	**TK004_Sk**—Protected Bird Area (PBA)
	Polygon	**TK005_Sk**—Site of Community Importance (SCI)
	Polygon	**TK007_Sk**—Ramsar site

(continued)

Table 5.5 (continued)

Component	Type	**Thematic layer—CODE in system** and explanation
	Polygon/	**USES_Sk**—Territorial System of Ecological Stability (TSES)
	Point	**Naucchod_Sk**—nature trail
	Polygon	**Lesyucel_Sk**—purposed forest
	Point	**Pamiatky_Sk**—cultural monument
SEF of protection of natural resources	Polygon	**Najkvalp_Sk**—highest quality soil (I–IV. Bonity)
	Polygon	**TW001_Sk**—Protected water management area (PWMA)
	Point	**RW004_Sk**—water source
	Polygon	**TW003_Sk**—protective zone of water source
	Polygon	**TW004_Sk**—catchment of water course
	Point	**Minvody_Sk**—source of mineral water
SEF related to infrastructure	Polygon	**Cestyop_Sk**—protective zone of roads
	Polygon	**Zelezniceop_Sk**—protective zone of railways
	Polygon	**Elektrovodyop_Sk**—protective zone of electrical cables
	Polygon	**Produktovodyop_Sk**—protective zone of pipelines
Environment	Point	**Zdrojezo_Sk**—sources of air pollution
	Polygon	**Znecisto_Sk**—air pollution
	Line	**Znecistvt_Sk**—pollution of water courses
	Polygon	**Kontampzv_Sk**—contamination of ground water (degree of contamination $Cd > 2$)
	Polygon	**Kontamp_Sk**—contamination of soil
	Polygon	**Poskveget_Sk**—damage to vegetation
	Point	**Envirozataze_Sk**—environmental burden
	Polygon	**Envirozatazeop_Sk**—protective zone of environmental burden
	Line	**Intenzdop_Sk**—traffic intensity
Territorial units	Polygon	**TU001_Sk**—cadastral territory and municipality
	Point	**COV_Sk**—wastewater treatment plant
	Point	**Kanalizacia_Sk**—sewer
	Point	**Skladky_Sk**—landfill
	Point	**Vodovody_Sk**—water pipe
	Polygon	**Mikroregiony_Sk**—microregion
	Polygon	**Pôsobnostorg_Sk**—territorial area of state nature protection

Table 5.6 Information layers on all levels—selected secion (exúlanation of the content in Tables 5.1–5.5)

Component	Type	Thematic layer	Attribute, indicator	Dimension	Values of attribute	Characteristics
Relief	Raster	**MORFO**—morfometricsa	**Hillshade**—shade of relief	Number		Visual interpretation of expected shade of relief
	Raster		**Heights**—digital terrain model	[m n. m.]	(*Numeric* data)	Altitude above sea level of terrain
	Raster		**Heights_smt**—digital terrain model (smoothed)	[m n. m.]	(Numeric data)	Adjusted altitude above sea level of terrain
	Raster		**Slope**—slope	[°]	(Numeric data)	Slope value
	Raster		**Aspect**—orientation of relief	[°]	(Numeric data)	Value: azimuth
	Raster		**Curv_profile**—normal curavture	Number	(Numeric data)	Negative values: convex Positive values: concave
	Raster		**Curv_plan**—horizontal curvature	Number	(Numeric data)	Negative values: convex Positive values: concave
	Raster		**Flowlength**—length of slope	[m]	(Numeric data)	length of slope in metres from ridge to evaluated pixel
	Raster		**Flowacc_d8**—contributing area, microcatchment (D8)	[m^2]	(Numeric data)	Size of area above evaluated pixel using algorithm D8
	Raster		**Flowacc_dinf**—contributing area, microcatchment (D-infinite)	[m^2]	(Numeric data)	Size of area above evaluated pixel using algorithm D-infinite
	Raster		**Radiation**—sun power	[$Wh.m^{-2}$]	(Numeric data)	Sun power (direct + scattered) for 2010
	Raster		**Radiation_dur**—period of sunlight	[h]	(Numeric data)	Period of duration of direct sun for 2010

(continued)

Table 5.6 (continued)

Component	Type	Thematic layer	Attribute, indicator	Dimension	Values of attribute	Characteristics
	Line	**CA010_Sk**—contour	**CA010_Sk**—type of contour	Text	• Main contour	
					• Basic contour	
					• Additional contour	
					• Auxiliary contour	
			VYSKA—altitude m a.s.l. of contour	[m n. m.]	(Numeric data)	
			Shape_Length—length of section of contour	[m]	(Numeric data)	
	Point	**CA030_Sk**—high point	**NAZOV**—name of high point (peak)	Text	Name	
			VYSKA—altitude a.s.l of point	[m n. m.]	(Numeric data)	
	Polygon	Stated indicators are in the layer **KEK_Sk**	**NM_VYSKA**—average altitude m a.s.l. of LEC	[m n. m.]	(Numeric data)	
			SKLON—category of average slope of LEC	[°]	(Numeric data)	
			MORFO_POL—morphographic-position type of relief of LEC		Name	
			EXPOZICIA—predominant orientation of LEC	[°]	(Numeric data)	
			N_FORMA—normal shape of relief of LEC	Number	(Numeric data)	

(continued)

Table 5.6 (continued)

Component	Type	Thematic layer	Attribute, indicator	Dimension	Values of attribute	Characteristics
			H_FORMA—horizontal shape of relief LEC	Number	(Numeric data)	
Geological substrate	Polygon	**PG001_Sk**—site of geological unit	**GSK**—type of geological-substrate-hydrological complex	Text	Domains are stated for indicators of GSK layer KEK_Sk	
			Shape_Length—circumference of site of geological unit	[m]	(Numeric data)	
			Shape_Area—area of site of geological unit	[m^2]	(Numeric data)	
	Polygon	Stated indicators are in the layer KEK_Sk	**HG_INDEX**—hydrogeological types of sediments, weatherings and rocks			
			HG_T—coefficient of transmissivity of hydrogeological types of sediments, weatherings and rocks			
			HG_K—coefficient of filtration of hydrogeological types of sediments, weatherings and rocks			
			HG_SV—free capacity of hydrogeological types of sediments, weatherings and rocks			
Soil	Polygon	**PP001_Sk**—area of soil unit	**PODNY_TYP**—soil subtype	Text	Domains are stated under indicator PODNY_TYP of layer KEK_Sk	

(continued)

Table 5.6 (continued)

Component	Type	Thematic layer	Attribute, indicator	Dimension	Values of attribute	Characteristics
			Shape_Length—circumference of site of soil unit	[m]	(Numeric data)	
			Shape_Area—area of site of soil unit	[m^2]	(Numeric data)	
	Polygon	Stated indicators are in the layer KEK_Sk	**PODNY_DRUH**—soil texture (grain size)			
			HLBKA_PODY—soil depth			
			SKELET—skeletality of soil			
Water	Polygon	**PW001_Sk**—detailed catchment	**TOK**—name of watercourse of relevant catchment	Text	Name	
			HCP—hydrological number of catchment	Viacmiestny kód		
			PROFIL—section of course to which the catchment relates	Text		
			X—x coordinate of the mouth of the catchment in reference system S-JTSK	[m]	(Numeric data)	
			Y—y coordinate of the mouth of the catchment in reference system S-JTSK	[m]	(Numeric data)	
			Shape_Length—circumference of catchment (length of watershed of catchment)	[m]	(Numeric data)	
			Shape_Area—area of catchment	[m^2]	(Numeric data)	
			CTP—characteristics of shape of catchment (ratio of area to square of length of catchment)	Number		Shape to ideal shape

Table 5.7 List of topographical layers (Geodatabáza db2.gdb) relating to thematic layer RL001_Sk—Areas of land cover

Group of element of Current Landscape Structure CLS	Type	Code of layer in system	Characteristics of detail of element of CLS
Form of land use	Polygon	**les**	Forest
	Polygon	**priesek**	Section
	Polygon	**mociar**	Swamp, salt-marsh
	Point	**kroviny_b**	Scrub, thicket
	Polygon	**kroviny_p**	Scrub, thicket
	Polygon	**trava**	Grass, grass and shrub vegetation
	Polygon	**luka**	Meadow, pasture
	Polygon	**ov_sad**	Fruit orchard, garden
	Polygon	**vinica**	Vineyard
	Polygon	**orna_poda**	Arable land
	Point	**skala_b**	Rocks
	Polygon	**skala_p**	Rocks
	Polygon	**nevyuz_plocha**	Site without typical use
	Polygon	**mrtv_pod**	Infertile land
Non-forest greenery, small-area, point and line	Line	**zivy_plot**	Hedge, tree line, continuous strip of shrubs and trees
	Point	**strom**	Tree
Sites of urban vegetation	Polygon	**zelen**	Greenery
Sites of sport and free time	Polygon	**ihrisko**	Playing area
	Polygon	**kurt**	Court, tennis court
	Line	**sport_draha**	Sports track
	Polygon	**streln**	Shooting range
	Polygon	**tribuna**	Grandstand
	Polygon	**bazen**	Pool, fire water
	Line	**lanovka**	Cable car, ski life, suspended cable route
	Polygon	**amfiteater**	Amphitheatre

(continued)

Table 5.7 (continued)

Group of element of Current Landscape Structure CLS	Type	Code of layer in system	Characteristics of detail of element of CLS
	Line	**autokino**	Drive-in cinema
	Line	**atrakcia**	Attraction in fun park or water park
	Point	**pomnik_b**	Monument, memorial, cairn
	Polygon	**pomnik_p**	Monument, memorial, cairn
	Point	**fontana_b**	Fountain
	Polygon	**fontana_p**	Fountain
	Point	**kriz**	Shrine, cross
Non-connected buildings, structures	Polygon	**budova**	Building
	Point	**veza_b**	Tower
	Polygon	**veza_p**	Tower
	Line	**opevnenie_l**	Fortification, wall, bulwark
	Polygon	**opevnenie_p**	Fortification, wall, bulwark
	Polygon	**chatrc**	Shack, hut, cabin
	Point	**osvetl_zariad**	Lighting equipment
	Line	**plot**	Fence
Industrial and commercial sites	Point	**komin_b**	Chimney
	Polygon	**komin_p**	Chimney
	Polygon	**sklad**	Store
	Polygon	**obilne_silo**	Grain silo, granary
	Point	**trznica**	Marketplace
Mining sites and structures	Polygon	**stena_lomu**	Wall of quarry
	Point	**taz_veza_b**	Mining tower
	Polygon	**taz_veza_p**	Mining tower
Sites and structures of land transport	Line	**ces_usek**	Road section
	Polygon	**plocha_lin**	Area around line object

(continued)

Table 5.7 (continued)

Group of element of Current Landscape Structure CLS	Type	Code of layer in system	Characteristics of detail of element of CLS
	Polygon	**parkovisko**	Parking
	Line	**chodnik**	Path
	Point	**priepust**	Sluice
	Line	**zel_usek**	Section of railway
	Polygon	**rampa**	Ramp (place for loading/unloading material in the railway)
	Polygon	**most_kon**	Bridge structure
	Line	**mostny_obluk**	Bridge arch
	Point	**pylon**	Pylon tower, pylon
	Point	**brána**	Gate, barrier
Sites and structures of boat transport	Line	**nabrez**	Bank, berth
	Polygon	**molo**	Pier
	Point	**brod_b**	Ford
	Polygon	**brod_p**	Ford
Sites of aviation transport	Polygon	**vp_draha**	Runway
	Polygon	**prist_hel**	Landing pad for helicopters and VTOL aeroplanes
Pipelines, energy lines, telecommunication	Line	**potrubie**	Pipeline
	Polygon	**vodojem_p**	Water tank
	Point	**vodojem_b**	Water tank
	Polygon	**tank_plyn**	Tank for storing liquids (apart from drinking water) and gases
	Point	**stlp_lan**	Column of hanging track, cable car, supporting bridge
	Line	**dopravnik**	Conveyor

(continued)

Table 5.7 (continued)

Group of element of Current Landscape Structure CLS	Type	Code of layer in system	Characteristics of detail of element of CLS
	Polygon	**nasypnik**	Hopper, container of bulk materials
	Line	**nadz_elekt**	Electric power lines
	Point	**transf**	Transformer, switch gear
	Point	**stlp_ele**	Column, post for electric lines
	Polygon	**radar_p**	Radar transmitter, receiver
	Point	**radar_b**	Radar transmitter, receiver
Water management works and structures	Line	**hradza_l**	Dam, weir, sluice
	Polygon	**hradza_p**	Dam, weir, sluice
	Line	**kor_hradz_l**	Crown of dam
	Polygon	**kor_hradz_p**	Crown of dam
	Line	**vod_pad_l**	Waterfall
	Point	**ponor_rieky**	River flows into ground
Waste management structures	Polygon	**skladka_mat**	Landfill of material
	Polygon	**sedim_nadrz**	Sedimentation tank, sewerage pond
	Line	**nasyp_l**	Dump, protective dam
	Polygon	**nasyp_p**	Dump, protective dam
Other structures and areas	Line	**zarez_l**	Section
	Polygon	**zarez_p**	Section
	Point	**ostat_ob_b**	Other structure
	Line	**ostat_ob_l**	Other structure
	Polygon	**ostat_ob_p**	Other structure
	Point	**neident_ob_b**	Unidentified structure
	Line	**neident_ob_l**	Unidentified structure
	Polygon	**neident_ob_p**	Unidentified structure

Table 5.8 List of indicators related to thematic layer TU001_Sk—Cadastral area of municipality

Name of file xls (in slovak)	Code	Name of characteristic (translated) [unit]
POHYB OBYVATEĽSTVA	U04090	Number and balance of migration total [persons]
ÚMRTNOSŤ1	U04161	Death total [persons]
ÚMRTNOSŤ2	U04212	Death according to illness [persons]
OBYVATEĽSTVO SPOLU	U25010	Total residents [persons]
OBYVATEĽSTVO PODĽA VEKOVÝCH SKUPÍN	U25391	Residents by age group [persons]
VZDELANIE	U25421	Highest level of education [persons]presently]
EKONOMICKY AKTÍVNE OSOBY	U25470	Working persons—with steady work
EKON. AKTÍVNE OSOBY PODĽA OKEC	U25512	Working persons according to sectors [persons]
ROZVODNÉ SIETE	U03080	Public water pipe [yes/no]
ZRÁŽKY	U21050	Total precipitation per year [millimetres]
SNEHOVÁ POKRÝVKA	U21115	Days per year with snow cover [number in units]
CELKOVÁ PLOCHA OBCE	U14010	Total area of village—town [m^2]
ZELEŇ	U23030	Public greenery total
KNIŽNICE	U12010	Public libraries [Number]

5.2 Spatial Projection of Database onto Maps

All defined elements of geocomplexes have been processed on maps, both analytically, using the topical model of geosystems, as well as synthetically using the choric model. This means that the complete spatial information system consists of a set of map layers processed by GIS technology, which enables the creation of all sorts of combinations of indicators of properties of elements of geocomplexes.

On the following pages we present some samples of sections from the maps of thematic layers: (Fig. 5.2, 5.3, 5.4, 5.5, 5.6, 5.7, 5.8, 5.9, 5.10, 5.11 and 5.12).

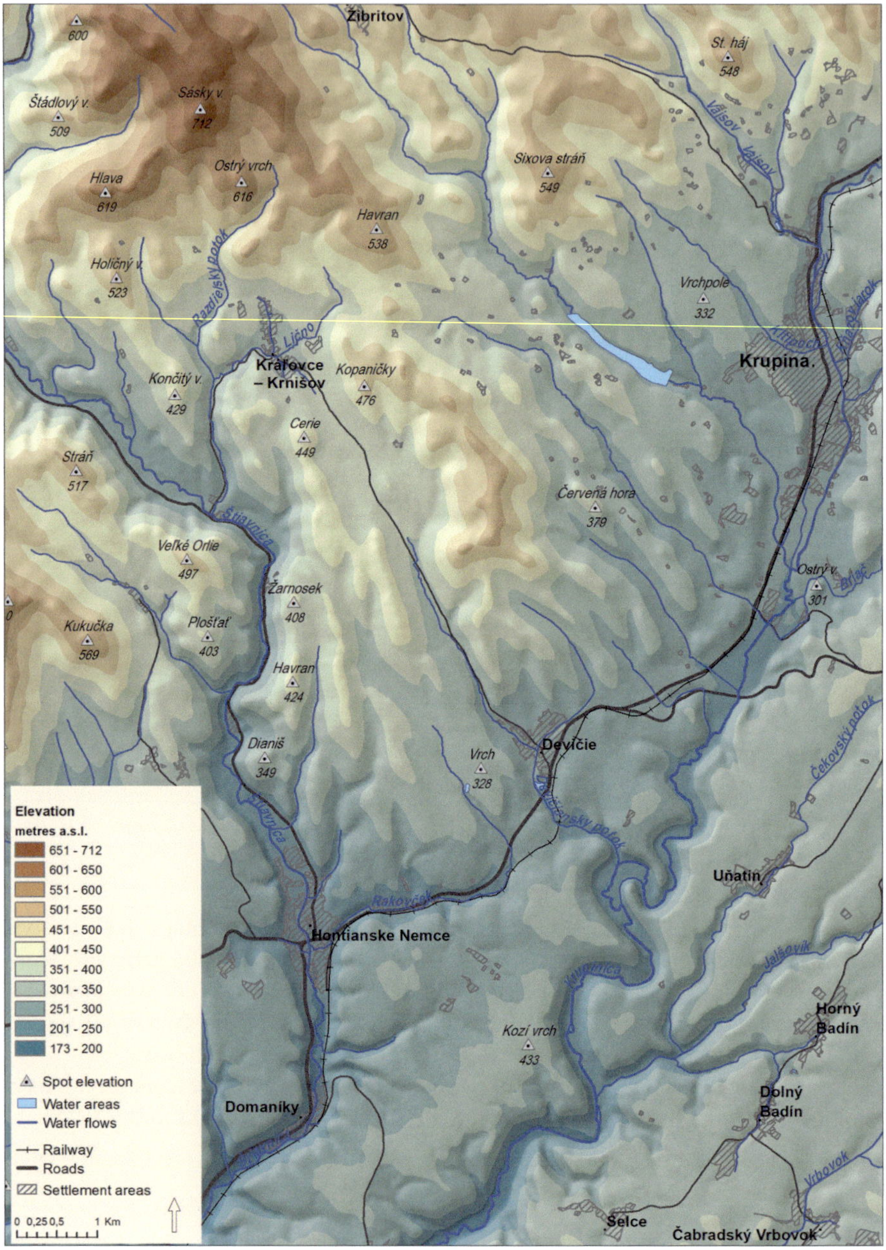

Fig. 5.2 Basic map—unified cartographic basis for all other thematic layers

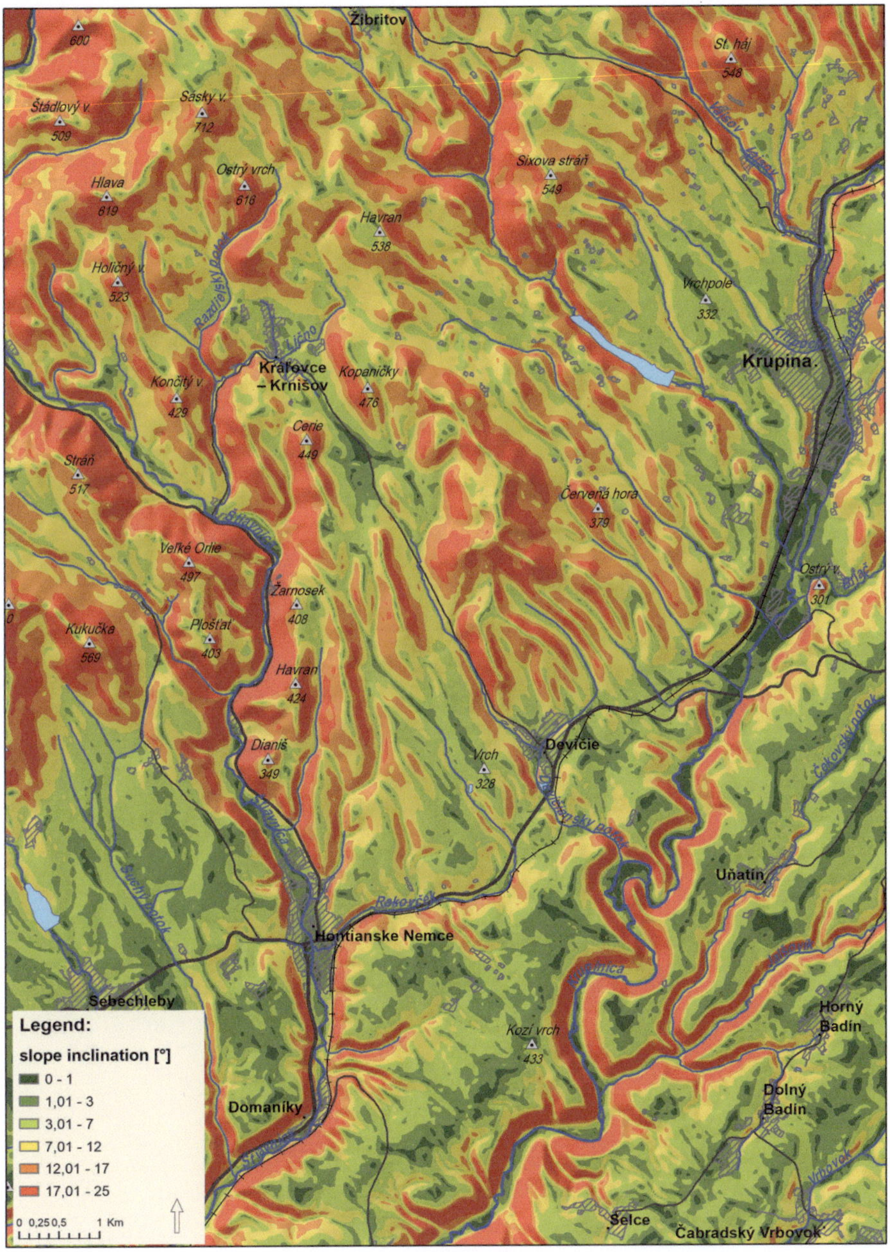

Fig. 5.3 Slope inclination. Thematic layer “SLOPE—slope inclination”. Values according to Tables 4.2, 5.2 and 5.6

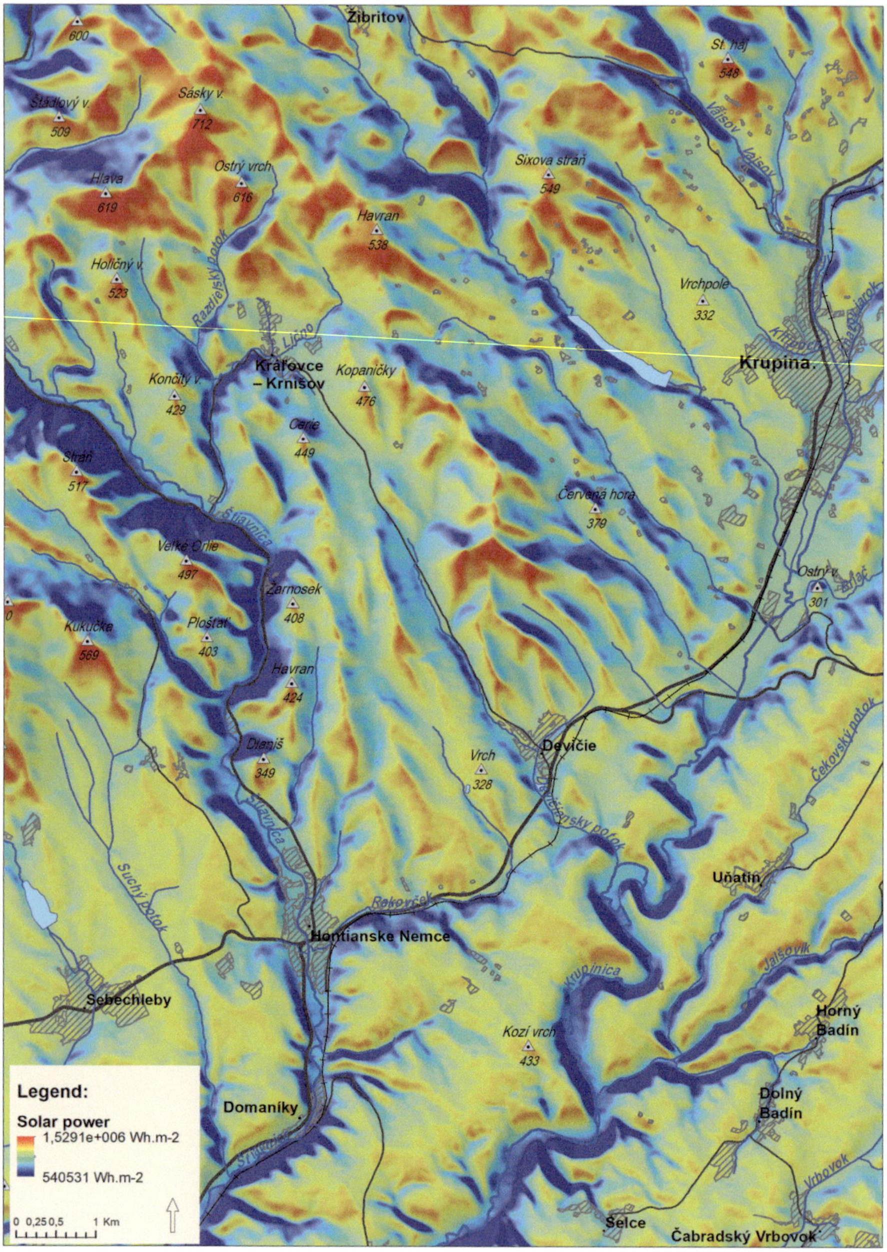

Fig. 5.4 Income of solar energy. Thematic layer "RADIATION—sun power". Values according to Tables 4.12, 5.2 and 5.6

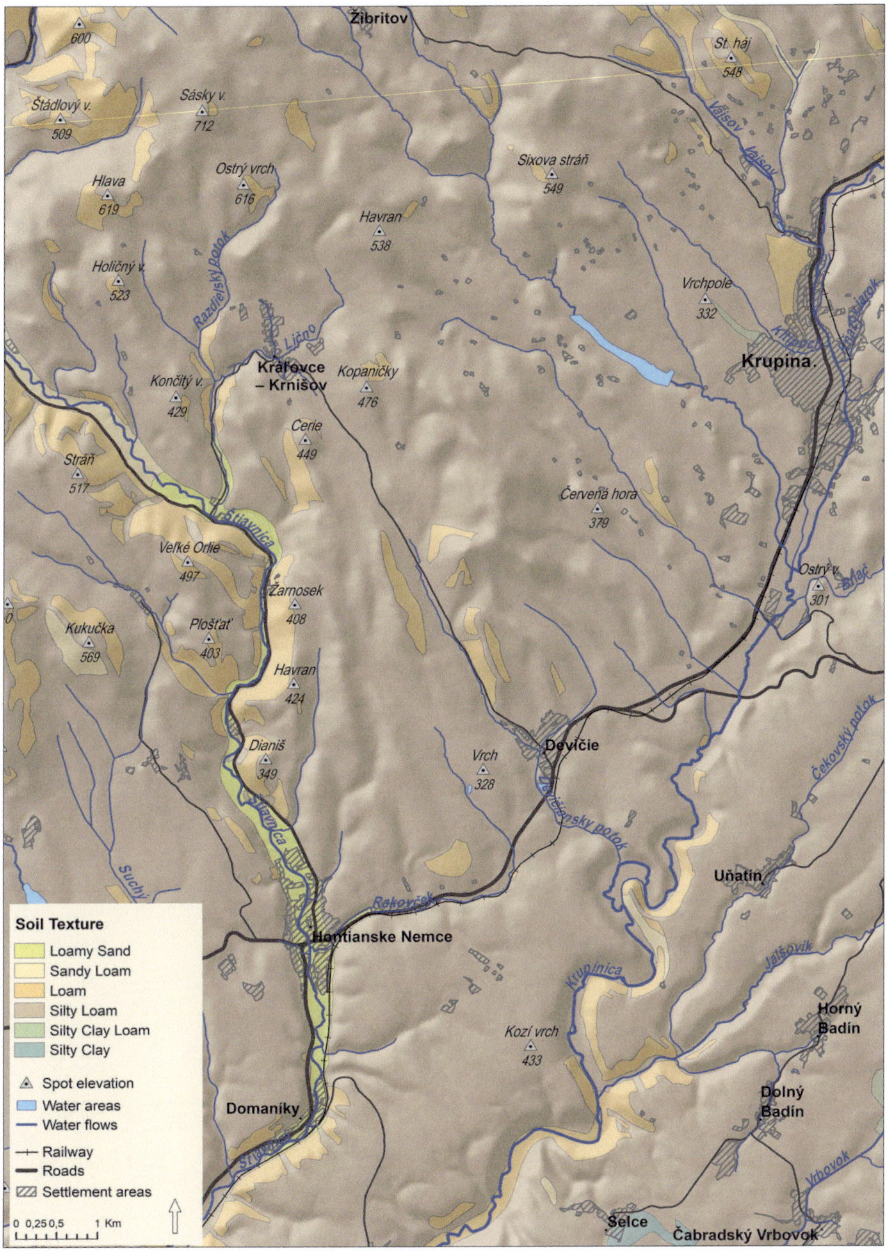

Fig. 5.5 Soil texture (grain size). Presented within the thematic layer "KEK_Sk—Landscape ecological complex" attribute "Podny_druh—soil texture (grain size)". Values according to Table 5.4

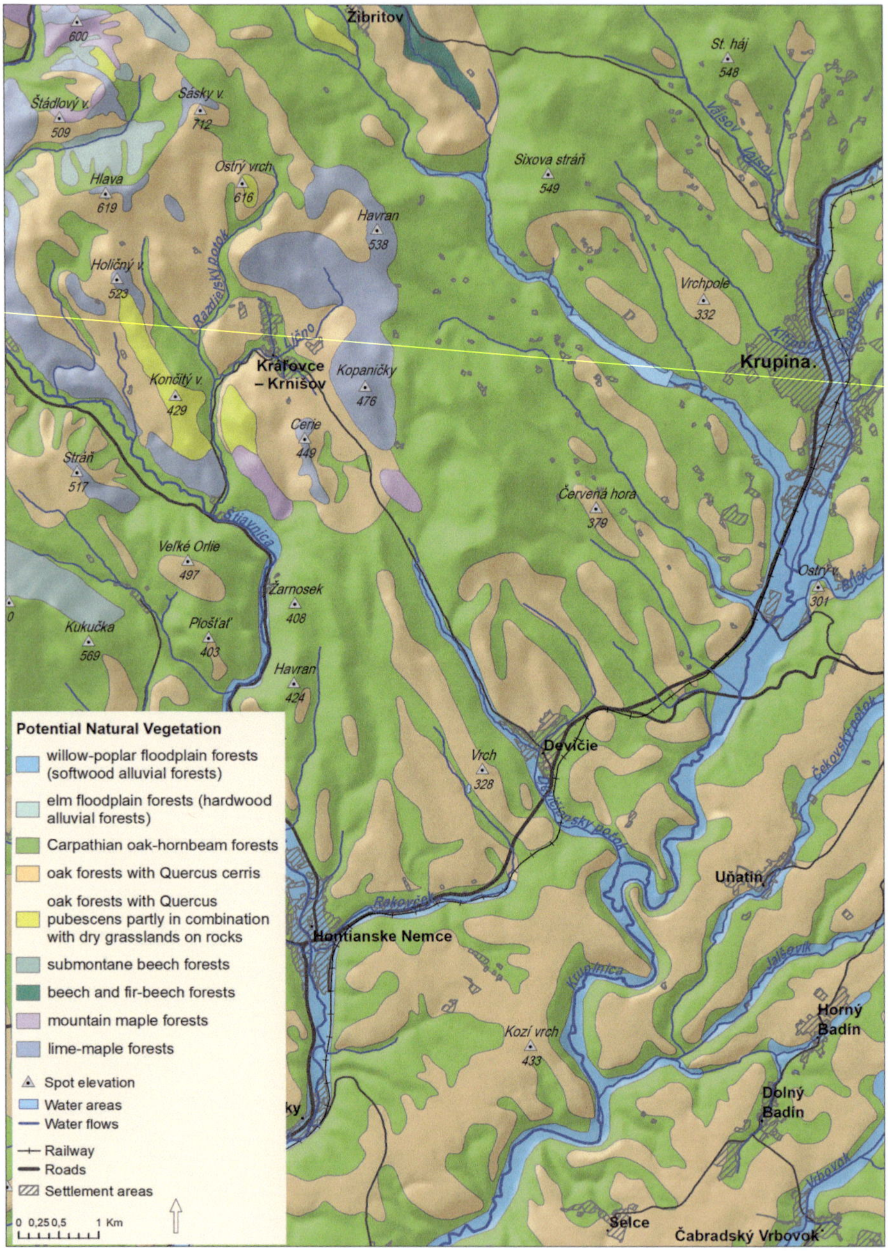

Fig. 5.6 Potential natural vegetation. Thematic layer "POTVEGET_Sk—potential vegetation". Values according to Table 5.3 (according to Maglocký 2002)

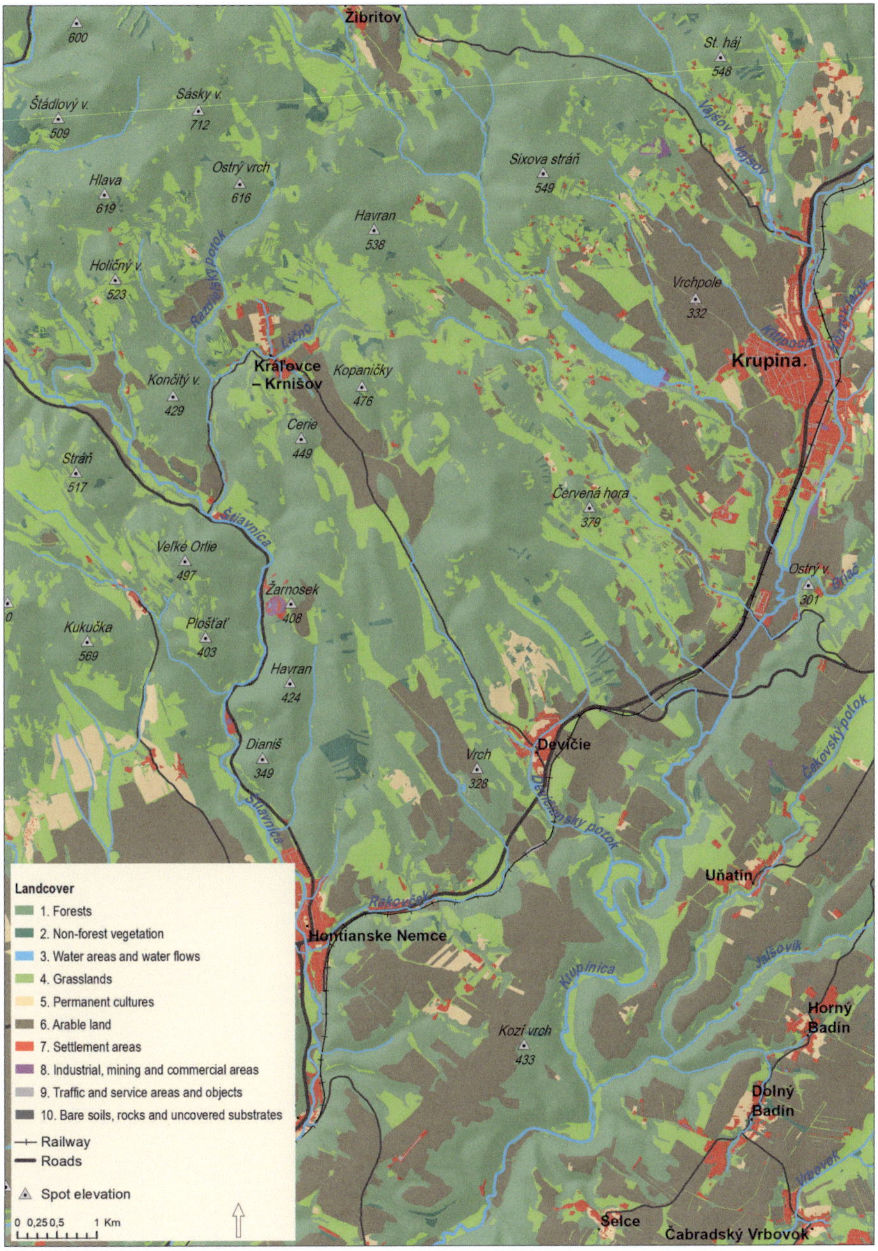

Fig. 5.7 Current landscape structure/land cover. Thematic layer "RL001_Sk—land cover". Values according to Tables 5.3 and 5.7. On map reduced information

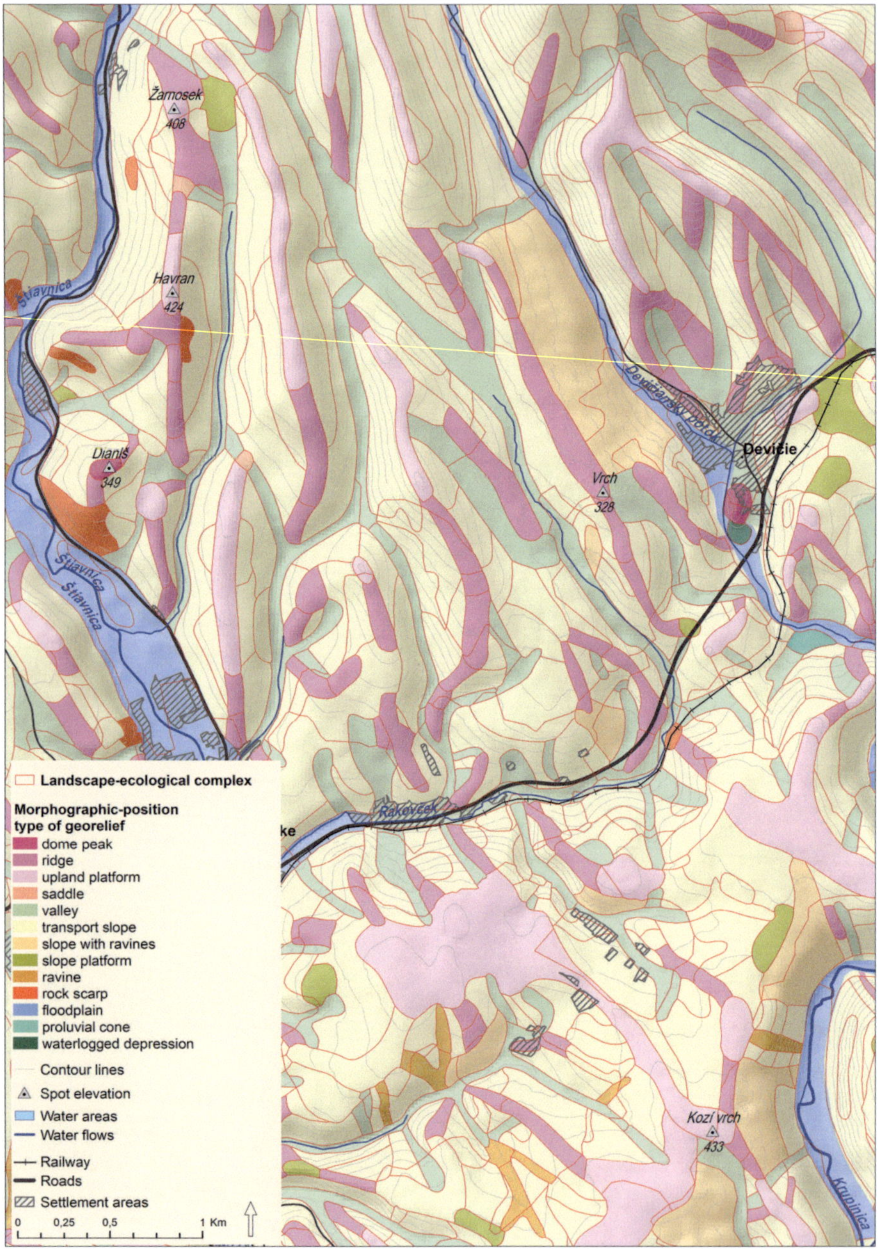

Fig. 5.8 Morphographic- positional types of the relief. Presented within the thematic layer "KEK_Sk—Landscape ecological complex" as attribute "Morfo_pol—morphographic-position type of relief of LEC". Values according to Table 5.4

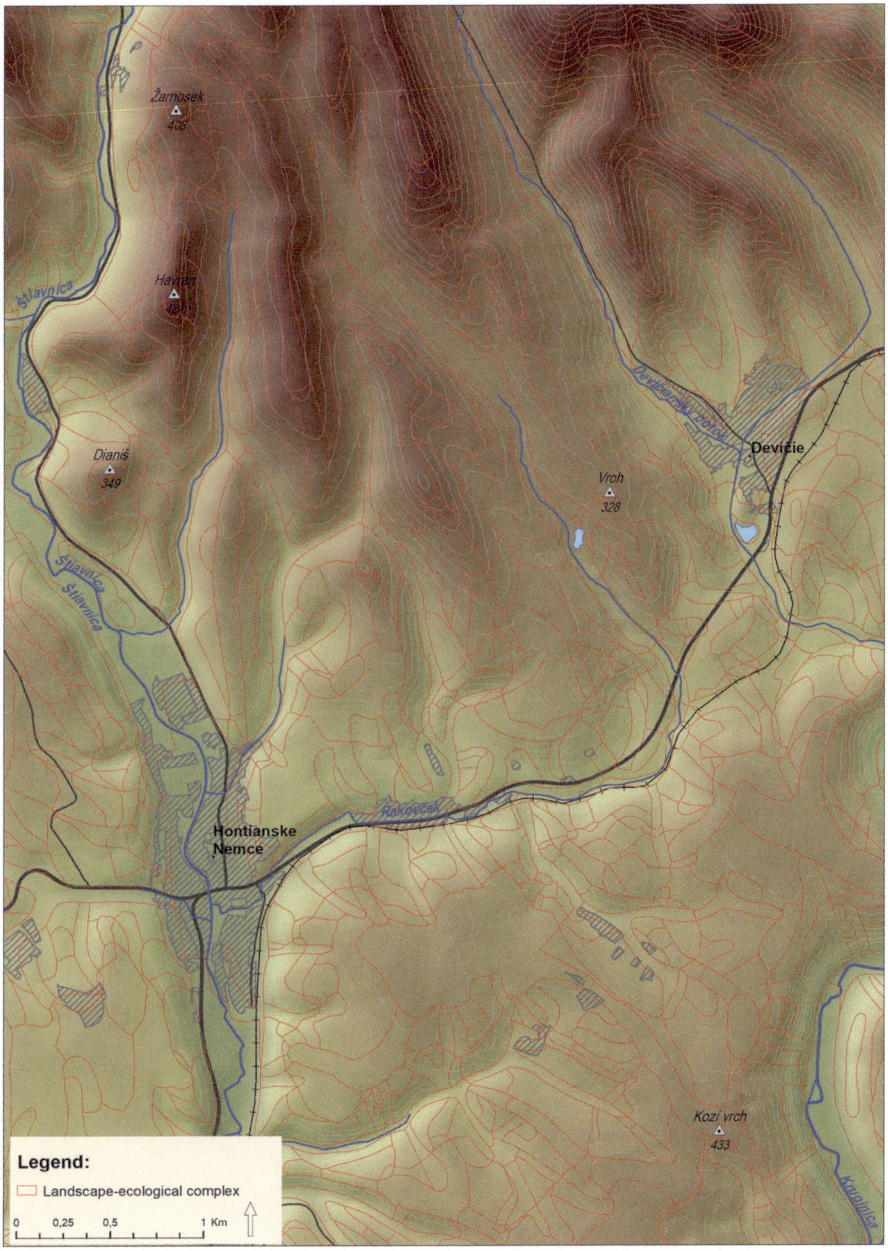

Fig. 5.9 Synthetic map of landscape-ecological complexes. Thematic layer "KEK_Sk—Landscape ecological complex" Each polygon is defined by values of all indicators listed in Tables 5.1 and 5.2

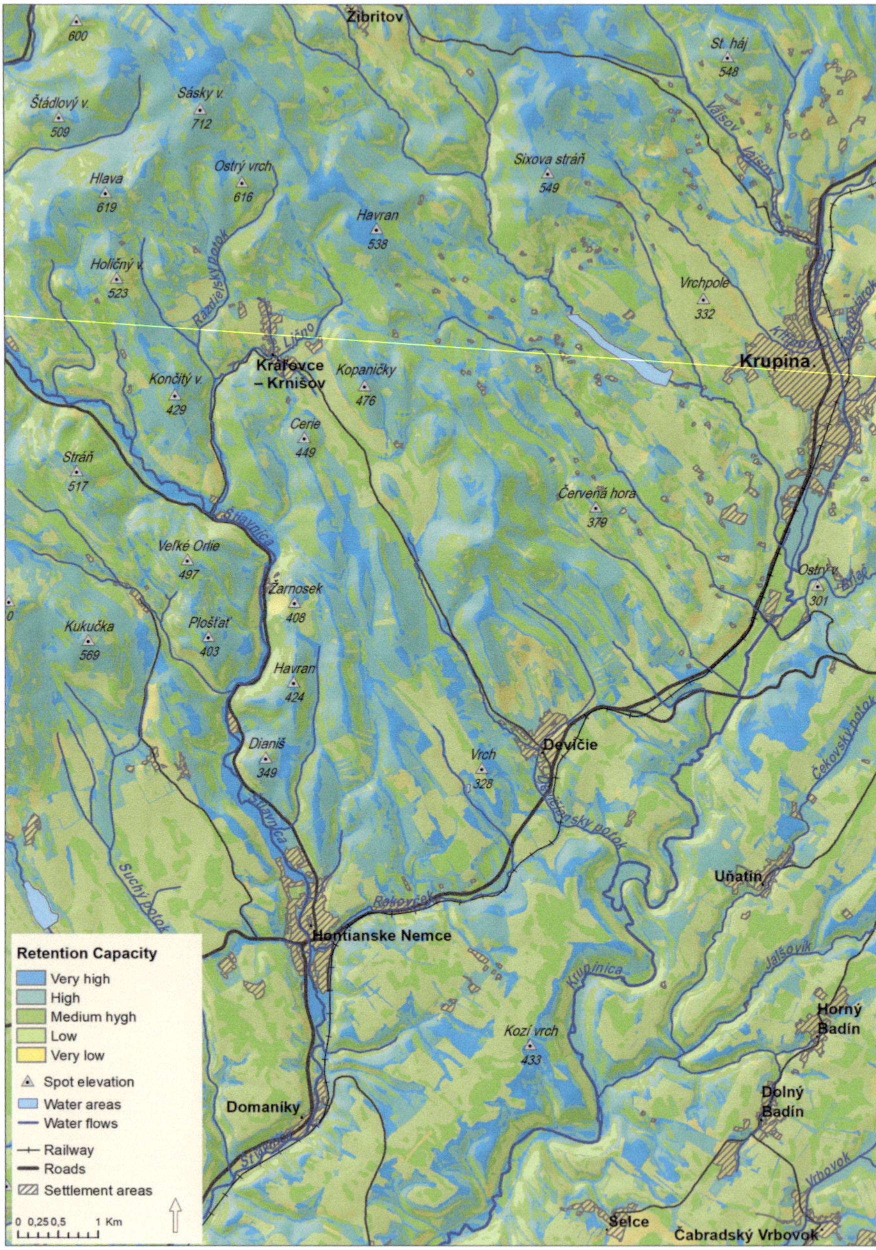

Fig. 5.10 Retention capacity of the landscape. Thematic layer "R—retention ability of territory". Values according to Table 5.4

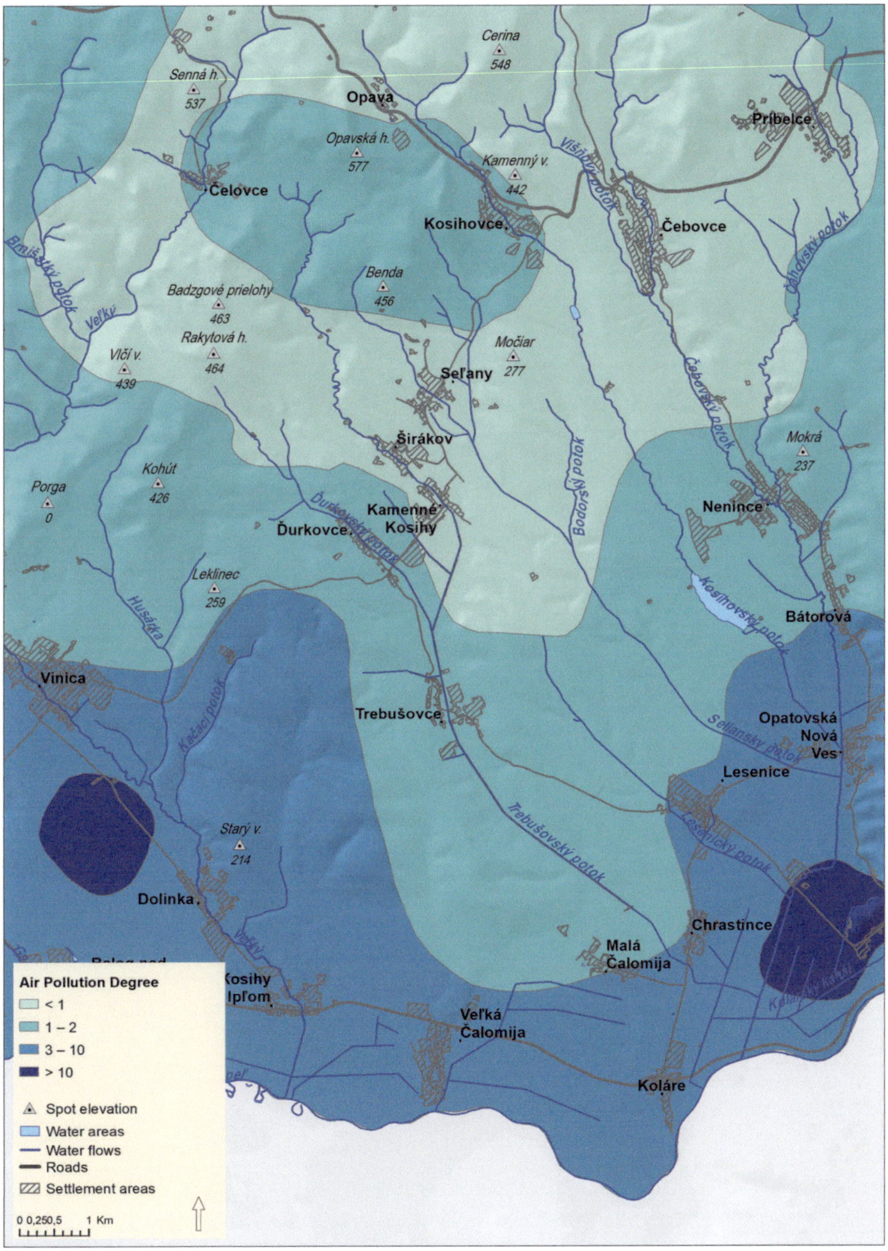

Fig. 5.11 Air pollution. Thematic layer "ZNECISTO_Sk—air pollution". Values according to Table 5.5

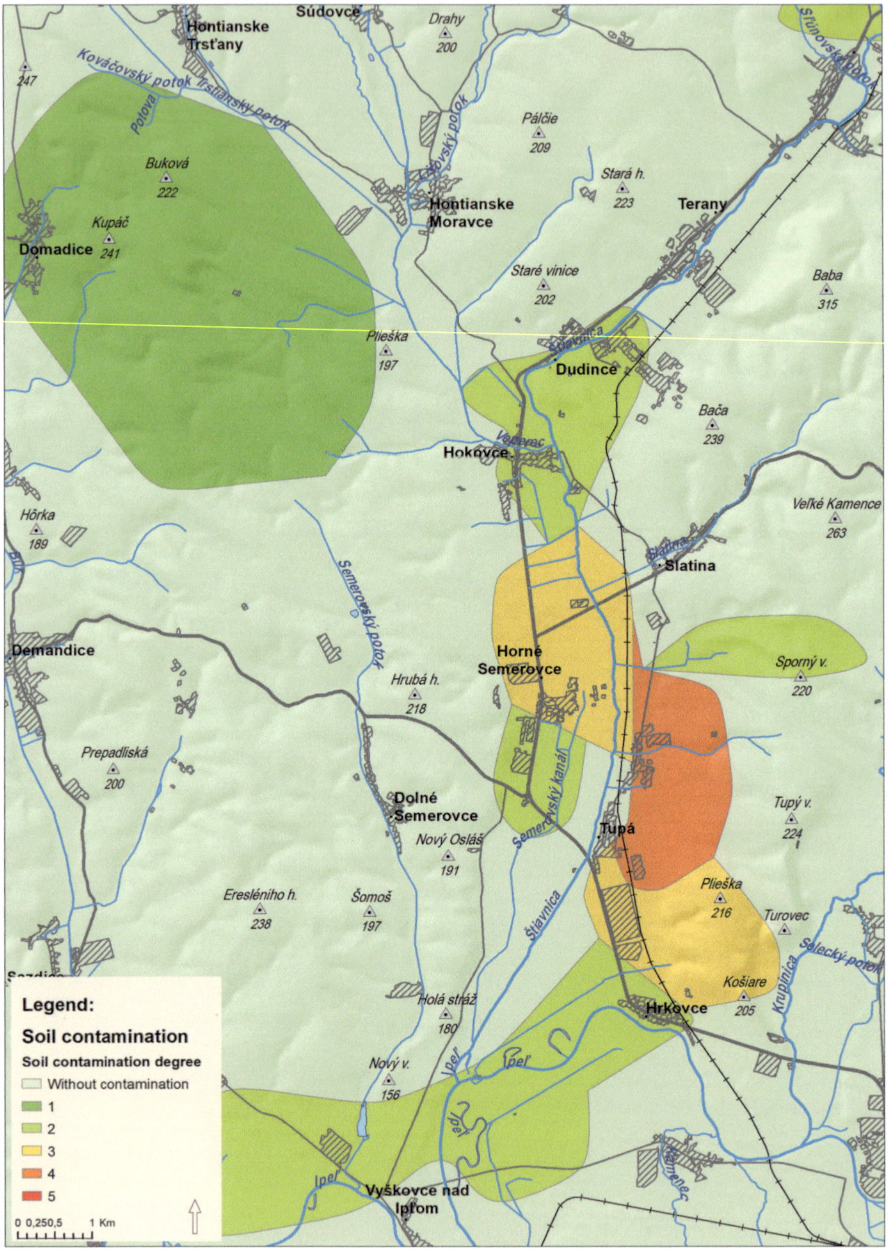

Fig. 5.12 Soil contamination. Thematic layer "KONTAMP_Sk—contamination of soil". Values according to Table 5.5

References

Collective (2008) Katalóg tried objektov ZB GIS. Úrad geodézie, kartografie a katastra SR, Bratislava, Topografický ústav, Banská Bystrica, pp 229

Conception (2006) Koncepcia tvorby, aktualizácie a správy ZB GIS na roky 2006–2010 (2006) Úrad geodézie, kartografie a katastra SR, Bratislava

Kocsis K, Agárdi N, Koczó F, Mezei G A, Nemerkényi Zs, Szabó R (2016) Generalizálás géppel, vagy kézzel? - módszertani összehasonlítás Magyarország Nemzeti Atlasza példáján. In: Pajtókné Tari I, Tóth A (eds) Magyar Földrajzi Napok: Absztraktkötet: VIII. Magyar Földrajzi Konferencia. Budapest: Magyar Földrajzi Társaság, p 103

Maglocký Š (2002) Potenciálna prirodzená vegetácia. In: Miklós L, Hrnčiarová T (eds) Atlas krajiny SR, MŽP SR, SAŽP Banská Bystrica, 1. vydanie, 114–115

Miklós L, Izakovičová Z, Kanka R, Ivanič B, Kočický D, Špinerová A, David S, Piscová V, Štefunková D, Oszlányi J, Ábrahámová A (2011) Geografický informačný systém povodia Ipľa: Katalóg GIS a výber máp. Bratislava: Ústav krajinnej ekológie SAV: Katedra UNESCO, Fakulta ekológie a environmentalistiky, Technická univerzita Zvolen, Esprit Banská Štiavnica, 143 pp

Miklós L, Diviaková A, Izakovičová Z (2011) Ekologické siete a územný systém ekologickej stability. Vydavateľstvo TU vo Zvolene, pp 141

Miklós L, Ivanič B, Kočický D (2011) Krajinnoekologická základňa integrovaného manažmentu povodia Ipľa. Digitálna databáza a tematické mapové vrstvy. Projekt HUSK 0801/2.1.2/0162 Vytvorenie jednotného monitoringu na báze priestorového informačného systému v povodí Ipľa. Banská Štiavnica, Esprit, spol. s r. o. (elektronický zdroj)

Miklós L, Németh R, Verrasztó Z (2014) Application of GIS in studying the drainage basin of the Ipoly River. Scientific Annals of the Danube Delta Institute, Tulcea, Romania 20:109–128

Németh R (2016) Társadalom és környezete a geoinformatikus szemével (Social Geographical Challenges and Search for Adequate Answers in EastCentral Europe of the 21st Century, Beregszász/Ukrajne. pp 600–607

Verrasztó Z (2010) Környezeti monitoring vizsgálatok az Ipoly vízgyűjtőjén (Tájökológiai Lapok. Gödöllő 8(3):532–561

Conclusion

One of the practical goals of the definition of the landscape as a geosystem is to provide a theoretical-methodical base for the creation of an integrated spatial information system usable for multiple purposes. It is a decisive and necessary condition for the development of integrated approaches to the landscape management. This can provide a unified spatial information base on the landscape accessible to all sectors. Understanding and implementing the measures to define and implement the target state of the environment is impossible without clear knowledge of the past and present states as well as changes to the landscape.

The cartographic visualization offers a special method of environmental impact assessment, because it makes it possible to assess the potential spatial consequences of the use of the environment. Cartographic visualization can show that the same or similar social impacts may not lead to identical environmental changes and consequences in other places.

Thematic maps represent the inner characteristics, structure and functions as well as the spatial structure of those natural and social phenomena that can be related to the surface of the Earth and show spatial distribution. This means that when the environmental sciences carry out their analysis they cannot neglect cartography, which has been using modern techniques. The cartographic approach uses the information technologies and offers great opportunities to understand the consequences of social impacts on the environment, to forecast the spatial distribution of impact mechanisms, and to explore transdisciplinary relationships. In this way, cartography meets the requirement of interoperability set by the directives of INSPIRE.

The whole chain of methodical steps starting with the theoretical aspects of geosystems through the characteristics of the indicators of geocomplexes up to the building the database and spatial information system for the model territory has been the framework of content of the presented book. We hope it has brought useful information for all readers.

L. Miklós et al., *Landscape as a Geosystem*,
https://doi.org/10.1007/978-3-319-94024-3

Cited Regulations and Conventions

European Landscape Convention, 2000, Council of Europe, Florence
Directive INSPIRE 2007/2/EC (Infrastructure for SPatial InfoRmation in Europe)
Act 50/1976 Coll. on territorial planning and building code (Building Act), as amended Act 237/2000 Coll.….
Act 543/2002 Coll. on nature and landscape protection, as amended
Act 3/2010 Coll. on the National Infrastructure of Spatial Information (NISI), as amended
Act 7/2010 Coll. on flood protection, as amended
Decree 24/2003 Coll. amending the Act on nature and landscape protection, as amended
Slovak Technical Norm (STN) 72 1001 Classification of soils and rock rocks (cancelled)
STN 72 1002 Classification of soils for road communications (cancelled)
STN 73 1001 Geotechnical constructions. Foundation of buildings (cancelled)

L. Miklós et al., *Landscape as a Geosystem*,
https://doi.org/10.1007/978-3-319-94024-3

MIX
Papier aus verantwortungsvollen Quellen
Paper from responsible sources
FSC® C105338
www.fsc.org

If you have any concerns about our products,
you can contact us on
ProductSafety@springernature.com

In case Publisher is established outside the EU,
the EU authorized representative is:
Springer Nature Customer Service Center GmbH
Europaplatz 3, 69115 Heidelberg, Germany

Printed by Libri Plureos GmbH
in Hamburg, Germany